Timer/Genera

Newnes Circuits Manual Series

Audio IC Circuits Manual R. M. Marston
CMOS Circuits Manual R. M. Marston
Electronic Alarm Circuits Manual R. M. Marston
Optoelectronics Circuits Manual R. M. Marston
Timer/Generator Circuits Manual R. M. Marston

Timer/Generator Circuits Manual

R. M. Marston

BPB PUBLICATIONS
B-14, CONNAUGHT PLACE, NEW DELHI-110001

FIRST INDIAN EDITION 1993, REPRINTED 2006
REPRINTED 2010

Distributors:

MICRO BOOK CENTRE
2, City Centre, CG Road,
Near Swastic Char Rasta,
AHMEDABAD-380009 Phone: 26421611

COMPUTER BOOK CENTRE
12, Shrungar Shopping Centre, M.G. Road,
BANGALORE-560001 Phone: 25587923, 25584641

MICRO BOOKS
Shanti Niketan Building, 8, Camac Street,
KOLKATTA-700017 Phone: 22826518, 22826519

BUSINESS PROMOTION BUREAU
8/1, Ritchie Street, Mount Road,
CHENNAI-600002 Phone: 28410796, 28550491

DECCAN AGENCIES
4-3-329, Bank Street,
HYDERABAD-500195 Phone: 24756400, 24756967

MICRO MEDIA
Shop No. 5, Mahendra Chambers, 150 D.N. Road,
Next to Capital Cinema V.T. (C.S.T.) Station,
MUMBAI-400001 Ph.: 220/8296, 22078297

BPB PUBLICATIONS
B-14, Connaught Place, **NEW DELHI-110001**
Phone: 23325760, 23723393, 23737742

INFO TECH
G-2, Sidhartha Building, 96 Nehru Place,
NEW DELHI-110019
Phone: 26438245, 26415092, 26234208

INFO TECH
Shop No. 2, F-38, South Extension Part-1
NEW DELHI-110049
Phone: 24691288, 24641941

BPB BOOK CENTRE
376, Old Lajpat Rai Market,
DELHI-110006 PHONE: 23861747

Copyright © R.M. Marston. All Rights Reserved.
No part or this publications may be reproduced, stored in a retrieval system, or transmitted in any fo
or by any means, electronic, mechanical, photocopying, recording, or otherwise, without the p
permission of the copyright owner.

All brand names and product names mentioned in this bok are trademarks or service marks of t
respective companies. Any omission or misuse (of any kind) of service marks or trademarks should no
regarded as intent to infringe on the property of others. The publisher recognizes and respects all m
used by companies, manufacturers, and developers as a means to distinguish their products.
THIS EDITION IS AUTHORIZED FOR SALE IN INDIAN SUB CONTINENTS ONLY.

Printed in India by arrangement with
BUTTERWORTH-HEINEMANN LTD.,ENGLAND

ISBN 81-7656-793-0

Published by Manish Jain for BPB Publications, B-14, Connaught Place, New Delhi-110 001 and
Printed by him at Akash Press, Delhi.

Contents

Preface		vii
1	Basic principles	1
2	Sine-wave generators	14
3	Square-wave generators	30
4	Pulse generator circuits	63
5	Timer IC generator circuits	87
6	Triangle and sawtooth generators	124
7	Multi-waveform generation	134
8	Waveform synthesizer ICs	142
9	Special waveform generators	164
10	Phase-locked loop circuits	182
11	Miscellaneous 555 circuits	226
Appendix Design charts		254
Index		265

Preface

This book is concerned mainly with waveform generator techniques and circuits. Waveform generators are used somewhere or other in most types of electronic equipment, and thus form one of the most widely used classes of circuit. They may be designed to produce outputs with sine, square, triangle, ramp, pulse, staircase, or a variety of other forms. The generators may produce modulated or unmodulated outputs, and the outputs may be of single or multiple form.

Waveform generator circuits may be built using transistors, op-amps, standard digital ICs, or dedicated waveform or 'function' generator ICs. One of the most popular ways of generating square and pulse waveforms is via so-called 'timer' ICs of the widely available and versatile '555' type, and many circuits of this type are included in this book.

The manual is divided into eleven chapters, and presents a total of over three hundred practical circuits, diagrams and tables. The opening chapter outlines basic principles and types of generator. Chapters 2 to 8 each deal with a specific type of waveform generator, and Chapter 9 deals with special waveform generator circuits. Chapter 10 takes an in-depth look at phase-locked loop circuits, and the final chapter deals with miscellaneous applications of the ubiquitous '555' timer type of IC. A special appendix presents a number of useful waveform-generator design charts, as an aid to those readers who wish to design or modify generator circuits to their own specifications.

The book is specifically aimed at the practical design engineer, technician and experimenter, but will be of equal interest to the electronics student and the amateur. It deals with its subject in an easy-to-read, down-to-earth, non-mathematical but very comprehensive manner. Each chapter starts off by explaining the basic principles of its subject and then goes on to present the reader with a wide range of practical circuit designs.

Throughout the volume, great emphasis is placed on practical 'user' information and circuitry, and the book abounds with useful circuits and data. Most of the ICs and other devices used in the practical circuits are modestly priced and readily available types, with universally recognized type numbers.

R. M. Marston

1 Basic principles

Electronics is primarily concerned with the business of signal or waveform processing and manipulation. This may involve the amplifying or shaping of one signal, or the mixing of two or more waveforms to give a complex modulated output, or the processing of a complex signal to extract its original components, or the use of one waveform to trigger a sequence of operations, etc. All these processes involve the use of waveform generators, which thus form a major class of circuit and may be designed to produce outputs with specific shapes (such as sine, square, or triangle), or to produce waveforms of exceptional purity or frequency stability, and may have single or multiple outputs, which may be modulated or unmodulated.

Specific waveforms can be generated directly, using, for example, oscillators or multivibrators, or they can be synthesized by using special *function generator* or phase-locked loop (PLL) techniques. Waveform generator circuits may be built using transistors, op-amps, standard digital ICs, or dedicated waveform or function generator ICs. This opening chapter looks at waveform basics and explains some of the techniques that are used in waveform generation.

Free-running waveforms

Electronic waveforms can be generated in either free-running or triggered form. Free-running types generate a waveform that completes each cycle in a period (P) and automatically repeats the generation process *ad infinitum*, repeating the cycles at a frequency (f) of $1/P$ Hz. Triggered types, on the other hand, generate a single waveform cycle on the arrival of each input trigger signal.

The four most widely used free-running waveforms are the sine, square, triangle and sawtooth types, and these are characterized by each having a unique harmonic structure, as shown in *Figure 1.1*. Thus, a *pure* **sine wave** generates a signal at its fundamental frequency only, and produces no harmonics (signals at precise multiples of the fundamental frequency), as shown in the diagram. All real sine waves are impure and generate unwanted harmonics. Sine wave purity is qualified by summing the strength of these unwanted harmonics and comparing them with those of the fundamental, to arrive at a final total harmonic distortion (THD) figure.

A *pure* **square wave** has perfect symmetry, switches between the high and low states in zero time, and generates an infinite number of odd harmonics with relative strengths directly related to their harmonic numbers. Thus, the seventh harmonic has a strength that is one-seventh of that of the fundamental, and so on.

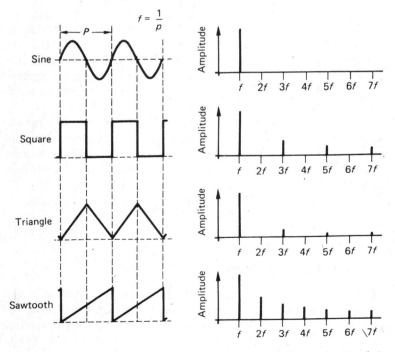

Figure 1.1 *The four most widely used free-running waveforms, showing their harmonic structures*

A *pure* **triangle wave** has perfect symmetry and linearity, but when analysed acts like an integrated squarewave. It produces only odd harmonics, but these are relatively weak and die away quite rapidly.

Finally, a *pure* **sawtooth** waveform rises with perfect linearity, switches between the high and low states in zero time, and generates an infinite number of even and odd harmonics, each with a strength proportional to the harmonic number, as shown in the diagram.

Triggered waveforms

Note from *Figure 1.1* that the shape of a free-running waveform cycle is quite independent of variations in frequency. Thus, if frequency is doubled, all horizontal waveform dimensions are simply halved, and the shape of each cycle is unaltered. Triggered waveforms give the opposite of this action, and the overall wave cycle shape varies with repetition frequency, as illustrated in *Figure 1.2*, which shows the two most popular triggered waveform types: the pulse and the sawtooth.

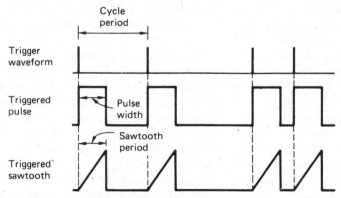

Figure 1.2   *Triggered pulse and sawtooth waveforms*

Thus, **triggered pulse** and **triggered sawtooth** waveforms are each characterized by the fact that their pulse or sawtooth width is absolutely constant, and is quite unaffected by variations in repetition period; the cycle wave shape thus varies with frequency, as shown.

Square-wave basics

Square waves can be generated either directly or by 'conversion' from an existing waveform. *Figure 1.3* illustrates the basic parameters of a

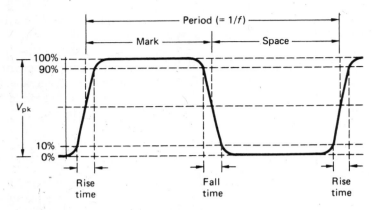

Figure 1.3 *Basic parameters of a square wave*

square wave; in each cycle the wave first switches from zero to some peak voltage value (V_{pk}) for a fixed period, and then switches low again for a second fixed period. The waveform takes a finite time to switch between states. The time it takes to rise from 10% to 90% of V_{pk} is known as its **rise time**, and that taken for it to drop from 90% to 10% of V_{pk} is known as its **fall time**.

Low-quality square waves have fairly long rise and fall times, and are easily produced via what are colloquially known as *squirt* generators. This type of waveform is useful in non-critical applications such as relay driving, LED flashing, sound generation, etc. High-quality square waves have very short rise and fall times, and are produced via so-called *clock* generators. This type of waveform is essential for correctly clocking fast-acting digital counter and divider ICs, etc.

In each square wave cycle the *high* part is known as its **mark** and the *low* part as its **space**. In a symmetrical square wave (such as *Figure 1.3*) the mark and space periods are equal and the waveform is said to have a 1:1 M/S ratio, or a 50% duty cycle (since the mark duration forms 50% of the total cycle period). Square waves *do not* have to be symmetrical, however, and their M/S ratios, etc., can be varied over a very wide range, as illustrated in *Figure 1.4*.

Note from *Figure 1.4* that the *mean* output voltage (V_{mean}) of each waveform, integrated over one complete cycle period, is equal to V_{pk} multiplied by the waveform's percentage duty cycle. Thus, if V_{pk} has a value of 10 V, the waveform (which has a 1:9 M/S ratio or 10% duty cycle) in *Figure 1.4(a)* will give a V_{mean} of 1 V; *Figure 1.4(b)* (which has a 1:1 M/S ratio or 50% duty cycle) will give a V_{mean} of 5 V; and *Figure*

Basic principles 5

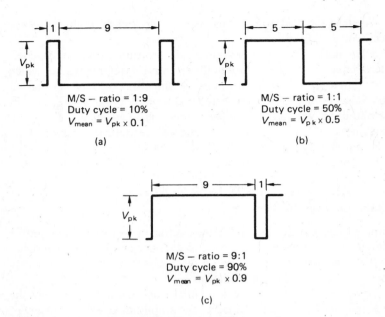

Figure 1.4 *Square waves with various mark/space ratios*

1.4(c) (which has a 9:1 M/S ratio or 90% duty cycle) will give a V_{mean} of 9 V. Thus, V_{mean} is fully variable via the M/S ratio or duty-cycle value.

Triangle or ramp waveforms

A pure triangle waveform has perfect symmetry, with equal rising and falling slope periods, as shown in *Figure 1.1*. Not all triangle waveforms are pure or symmetrical, however. Highly non-symmetrical waveforms are usually referred to as **ramp** generators, and *Figure 1.5* illustrates the typical range of waveforms that are available from a **variable-slope** ramp generator.

Special waveforms

Many special types of waveforms are also used in electronics.

Staircase waveforms rise between zero and some specific voltage value in a series of discrete time-related voltage steps and then switch to zero again. They are often used in curve tracers, etc.

Crystal-generated waveforms have exceptionally high frequency accuracy and stability.

White noise waveforms contain a full spectrum of randomly generated frequencies, all with randomly determined amplitudes but which have equal *power per bandwidth unit* when averaged over a reasonable unit of time. **Pink noise** is modified white noise, filtered so that its amplitudes have equal *voltage per bandwidth unit* when averaged over a reasonable unit of time. Both types of noise are useful in testing AF and RF amplifiers and in generating special sound effects.

Special types of waveform are described more fully in Chapter 9.

Waveform modulation

Simple basic waveforms can be used to transmit or carry analogue or digital data by using that data to modulate one or more characteristics of the basic waveforms. *Figure 1.6* illustrates four of the most widely used types of waveform modulation.

Free-running sine waves can be analogue modulated by using the analogue signal to modulate either the amplitude or the frequency of the sine wave, as in AM and FM systems respectively. These types of modulation are widely used in *voice* communication systems.

Sine waveforms can be used to carry simple two-state (binary) data and codes by using frequency shift keyed (FSK) modulation, in which the binary signal selects either a high- or a low-frequency tone. FSK is widely used in 'down-the-phone' data link systems.

Another way of transmitting analogue signals is to use them to modulate the pulse widths of triggered pulse generators, by using the

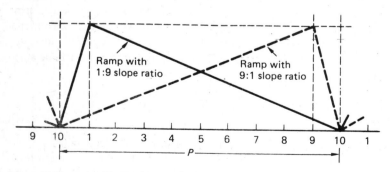

Figure 1.5 *Typical range of waveforms available from a variable-slope ramp generator*

Basic principles 7

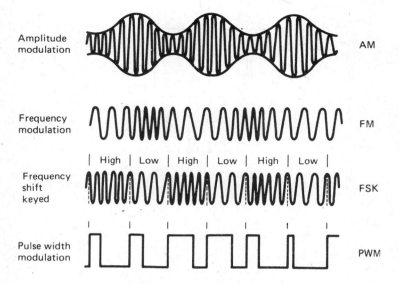

Figure 1.6 *Four widely used types of waveform modulation*

pulse width modulation (PWM) system, which is widely used in multichannel remote control systems, etc.

Waveform generating ICs

Most basic waveforms can be generated via simple transistor, op-amp, or digital IC circuits, or via a popular general-purpose IC such as the 555 timer, etc. Alternatively, where multiple or complex waveforms are wanted, these can often be generated via dedicated waveform generator or synthesizer ICs. The simplest of these takes the form of a voltage controlled oscillator (VCO), which generates simultaneous square and triangle outputs, has a voltage/frequency transfer graph similar to that shown in *Figure 1.7*, and can provide FM outputs by simply modulating the input control voltage.

More advanced VCO-based waveform generator ICs also provide a sine wave output, which is synthesized by rounding off the corners of the triangle waveform via a limiting or shaping circuit. Chapter 8 gives detailed descriptions of two such ICs, one of which also incorporates an AM-generating facility.

8 Timer/Generator Circuits Manual

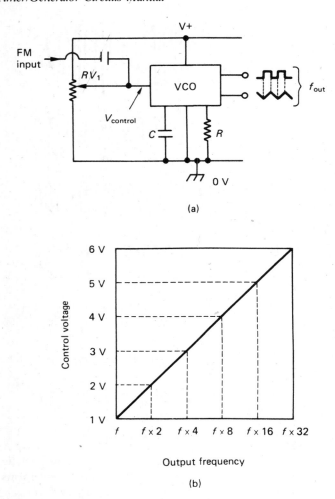

Figure 1.7 *Basic VCO circuit and typical voltage-to-frequency transfer graph*

Phase-locked loop basics

Some special waveforms can be generated by using a phase-locked loop (PLL). This is a circuit that automatically locks the frequency and phase of a VCO (f_o) to the mean frequency and phase of an external input reference signal (f_r). Such systems are useful in applications such as automatic frequency tracking, frequency multiplication, and frequency synthesis, etc. *Figure 1.8* shows the block diagram of a basic PLL system, which consists of a phase comparator, a loop filter and a VCO, and operates as follows.

The PLL's phase comparator receives the f_o and f_r signals, compares the phase and frequency of f_o with that of f_r, and generates a corresponding variable output error voltage which is then low-pass filtered and fed to the VCO's control input in such a way that any frequency or phase differences between f_o and f_r are progressively reduced until they fall to zero, at which point the loop is said to be *locked*.

Thus, if f_o is initially below f_r the phase comparator output goes positive and its resulting filtered voltage then makes the VCO frequency rise until both the frequency and phase of f_o precisely match those of f_r. If the VCO frequency rises above that of f_r the reverse action takes place, and the phase comparator output falls, reducing the VCO frequency until f_o again locks to f_r.

The low-pass filter is a vital part of the PLL system, and converts the phase detector's output into a smooth d.c. control voltage. Inevitably, it has a finite time constant, so PLL 'locking' is not instantaneous, and f_o locks to the mean value of f_r, rather than to its instantaneous value. This is useful if a clean output frequency is wanted from a noisy input signal.

Frequency multiplication

When the basic *Figure 1.8* PLL circuit is locked its output frequency equals the mean value of the input signal. *Figure 1.9* shows a *frequency multiplier* version of the PLL circuit, in which the output frequency is precisely ten times greater than the input signal. Here, a divide-by-ten counter is inserted in the feedback loop between the VCO output and the phase comparator input, and the phase comparator therefore locks to the output frequency of the counter (rather than to that of the VCO).

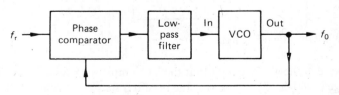

Figure 1.8 *Basic phase-locked loop (PLL) circuit*

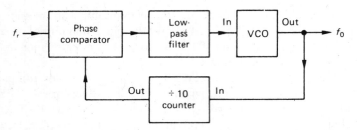

Figure 1.9 *Frequency multiplier circuit*

Thus, at lock the VCO frequency (f_o) is ten times greater than f_r. This circuit can be made to frequency-multiply by any desired number by simply using a counter (in the feedback loop) with an appropriate division ratio.

Frequency synthesis

Another useful application of the PLL is as a precision programmable frequency synthesizer. *Figure 1.10* shows one version of such a circuit. Note here that the phase comparator's reference input is a precision 1 kHz signal derived via a crystal oscillator, and that (as in *Figure 1.9)* a counter is wired into the feedback loop between the VCO output and the phase comparator input, but is externally programmable to give any whole-number division ratio value between × 100 and × 1000. Thus, this circuit can generate (synthesize) frequencies from 100 kHz to 1 MHz, in 1 kHz steps, with crystal accuracy and stability in each case.

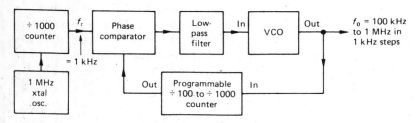

Figure 1.10 *Simple frequency synthesizer*

Note in the above circuit that the VCO must have a frequency *span* range of at least 10:1, to cover the required frequency range, and that the frequency *step* value corresponds to the 1 kHz external input value.

High-frequency synthesis

The programmable counter forms an essential part of all frequency synthesizers. In practice, these counters can usually handle maximum input frequencies of only a few megahertz, and the basic *Figure 1.10* circuit can thus not be used to directly synthesize high-frequency (above a few megahertz) signals. High-frequency PLL synthesizer circuits are available, however, and *Figures 1.11* to *1.13* show three different versions of this type of circuit.

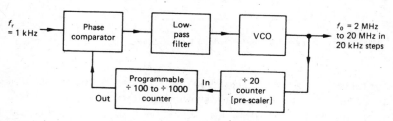

Figure 1.11 *Frequency synthesizer with prescaler*

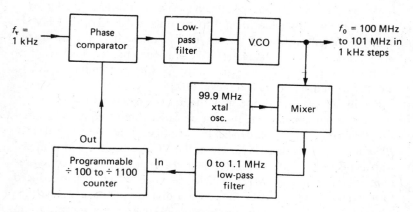

Figure 1.12 *High-frequency mixer-type synthesizer*

The *Figure 1.11* circuit uses a *pre-scaler* technique in which an additional *divide-by-X* high-frequency counter stage (the pre-scaler) is interposed between the VCO output and the programmable counter input, to enable the VCO to operate at a frequency X-times higher than the programmable counter stage. In the example shown the pre-scaler has a divide-by value of $\times 20$, enabling the synthesizer to cover

12 Timer/Generator Circuits Manual

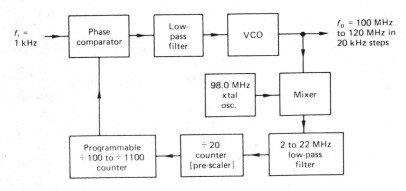

Figure 1.13 *Wide-range high-frequency synthesizer*

the range 2 MHz to 20 MHz in 900 discrete steps. A disadvantage of this technique is that it causes the synthesizer step value to increase by a ratio equal to the pre-scaler value, i.e., to $20 \times f_r$, or 20 kHz in this case.

The *Figure 1.12* circuit uses a *mixer* technique to synthesize frequencies in the range 100 MHz to 101 MHz in 1000 discrete steps of 1 kHz. In this case, the VCO output is mixed with a crystal-derived 99.9 MHz signal and then low-pass filtered to produce a 100 kHz to 1.1 MHz *difference* signal, which then passes into the phase-locked loop via the programmable counter stages. This technique enables the VCO frequency to be varied in steps equal to the f_r value, but limits the VCO's useful span range to only few megahertz.

Finally, *Figure 1.13* shows how the above *mixer* and *pre-scaler circuits* can be combined to make a wide-range, high-frequency synthesizer that can generate frequencies in the range 100 MHz to 120 MHz in 1000 discrete steps of 20 kHz. Here, the VCO output signal is mixed with a crystal-derived 98 MHz signal and then low-pass filtered to produce an output in the 2 MHz to 22 MHz range, which is then reduced to the 100 kHz to 1.1 MHz range via a divide-by-20 pre-scaler stage before being fed back into the phase-locked loop via the programmable counter stage. This type of synthesizer circuit gives excellent results.

The VCO

In high-frequency PLL synthesizers, the VCO is normally required to cover a very limited span range, and normally takes the form of a varicap-controlled transistor oscillator-plus-buffer circuit. In low-frequency synthesizers the VCO is normally required to cover a very wide span range, and normally takes the form of a special CMOS or bipolar oscillator. Some dedicated PLL ICs contain excellent wide-ranging VCOs which are outstandingly useful in their own rights. Among the best known of these ICs are the 4046B CMOS chip and the NE565 and NE567 PLL devices from Signetics, and all three of these devices are fully described in Chapter 10.

2 Sine-wave generators

The sine wave is the most fundamental and useful of all waveforms. Sine waves can be produced directly from suitable C–R or L–C oscillators, or can be synthesized via special waveform generator ICs. This chapter looks only at direct sine-wave generation via basic oscillator circuits. Sine-wave generation via special synthesizer ICs is described in detail in Chapter 8.

C–R oscillator circuits

Two basic requirements must be fulfilled to produce a simple sine-wave oscillator, as shown in *Figure 2.1*. First, the output of an amplifying device (A_1) must be fed back to its input via a frequency-selective network (A_2) in such a way that the sum of the amplifier and feedback-network phase-shifts equals 0° (or 360°) at the desired oscillation frequency, i.e., so that $x° + y° = 0°$ (or 360°). Thus, if a transistor amplifier gives 180° of phase shift between input and output, an additional 180° of phase shift must be introduced by a frequency-selective network connected between input and output to meet the first requirement of a sine-wave oscillator.

The second requirement for sine-wave oscillation is that the gain of the amplifying device must exactly counter the loss (attenuation) of the frequency-selective feedback network at the desired oscillation frequency, to give an overall system gain of precisely unity, e.g. $A_1 \times A_2 = 1$. Note that if the system gain is less than unity the circuit will not oscillate, and if greater than unity the system will be overdriven and will produce distorted (non-sinusoidal) waveforms.

The frequency-selective feedback network used in a sine-wave oscil-

Sine-wave generators 15

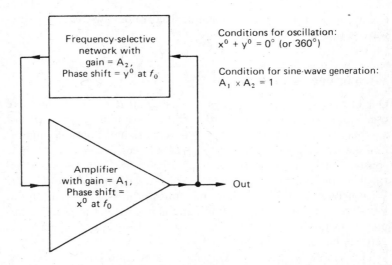

Figure 2.1 *Essential circuit and conditons needed for sine-wave generation*

lator usually consists of either a $C-R$ (capacitor–resistor) or an $L-C$ (inductor–capacitor) filter network. *Figure 2.2* shows the practical circuit of one of the crudest members of the sine wave oscillator family, the so-called **phase-shift oscillator**.

Here, the output (collector) signal of transistor amplifier Q_1 is fed back to its input (base) via a three-stage $C-R$ ladder network, essentially comprising C_1-R_1, C_2-R_2, and C_3-R_3. Each $C-R$ stage of the ladder produces a phase shift between its input and output terminals. The size of the shift depends on frequency and component values, and has a maximum value of 90°. The phase shift of the complete ladder network equals the sum of the shifts of each stage and, in *Figure 2.2* (in which $C_1 = C_2 = C_3 = C$, and $R_1 = R_2 = R_3 = R$), equals 180° at a frequency of $1/14\ C.R$. Since Q_1 itself produces a shift of 180°, the circuit actually oscillates at a frequency of about $1/14\ C.R$. Note that the three-stage ladder network gives an attenuation factor of about 29 at the oscillation frequency, and that a high-gain transistor must be used in the Q_1 position to compensate for this high circuit loss.

In use, the *Figure 2.2* circuit can be set up by carefully adjusting RV_1 until the circuit just goes into oscillation, thus producing a reasonably pure sine-wave output. In practice, oscillators of this type need frequent adjustment of the RV_1 gain control if good sine-wave purity is to be maintained, since the circuit has no inherent gain stability. Such circuits are useful, however, as simple fixed value low-frequency sine-wave generators.

Wien bridge oscillators

One of the best and easiest ways of making an *R–C*-based sine-wave oscillator is to connect a standard operational-amplifier (op-amp) and a frequency-selective Wien bridge *R–C* network in the basic configuration shown in *Figure 2.3*. Here, the frequency-sensitive Wien network is constructed from R_1–C_1 and R_2–C_2. Normally, the network is symmetrical, so that $C_1 = C_2 = C$, and $R_1 = R_2 = R$. The main feature of the Wien network is that the phase relationship of its output and input signals varies from $-90°$ to $+90°$, and is precisely $0°$ at a centre frequency (f_o) of $1/2\pi$ *CR*, or $1/6.28$ *CR*. At this centre frequency the network has a voltage gain of $\times 0.33$.

Thus, in *Figure 2.3*, the Wien network is connected between the output and the non-inverting input of the op-amp, so that the circuit gives zero overall phase shift at f_o, and the actual amplifier is given a voltage gain of $\times 3$ via feedback network R_3–R_4, to give the total system an overall gain of unity. The circuit thus provides the basic requirements for sine-wave oscillation. In practice, however, the ratios

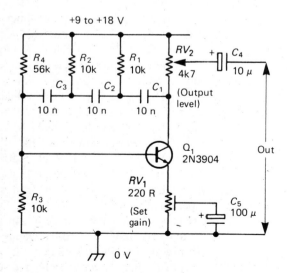

Figure 2.2   *800 Hz phase-shift oscillator circuit*

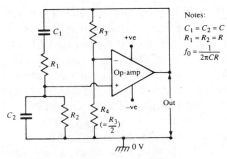

Figure 2.3 *Basic Wien bridge sine-wave oscillator*

of R_3–R_4 must be carefully adjusted to give the overall voltage gain of precisely unity that is necessary for low-distortion sine-wave generation.

The above circuit can easily be modified to give automatic gain adjustment and amplitude stability by replacing the passive R_3–R_4 gain-determining network with an active gain-control network that is sensitive to the amplitude of the output signal, so that gain decreases as the mean output amplitude increases, and vice versa. *Figures 2.4* to *2.8* show some practical versions of Wien bridge oscillators with automatic amplitude stabilization.

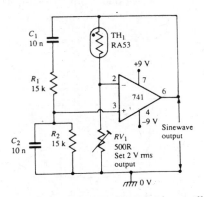

Figure 2.4 *Thermistor-stabilized 1 kHz Wien bridge oscillator*

Thermistor stabilization

In the 1 kHz fixed-frequency oscillator circuit of *Figure 2.4* the output amplitude is stabilized by an RA53 (or similar) negative-temperature-coefficient (NTC) thermistor. TH_1 and RV_1 form a gain-determining feedback network. The thermistor is heated by the mean power output of the op-amp, and at the desired output signal level has a resistance value double that of RV_1, thus giving the op-amp a gain of × 3 and the overall circuit a gain of unity. If the oscillator output amplitude starts to rise, TH_1 heats up and reduces its resistance, or vice versa, thereby automatically reducing the gain of the circuit and stabilizing the amplitude of the output signal.

An alternative method of thermistor stabilization is shown in *Figure 2.5*. In this case a low-current lamp is used as a positive-temperature-coefficient thermistor, and is placed in the lower part of the gain-determining feedback network. Thus, if the output amplitude increases, the lamp heats up and increases its resistance, thereby reducing the circuit gain and providing automatic amplitude stabilization. This circuit also shows how the Wien network can be modified by using a twin-gang pot to make the oscillator frequency variable over the range 150 Hz to 1.5 kHz, and how the sine-wave output amplitude can be made variable via RV_3.

In the *Figure 2.4 and 2.5* circuits, the pre-set pot should be adjusted to set the maximum mean output level to about 2 V r.m.s., and under this condition the sine wave has a total harmonic distortion of about 0.1%. Note that a slightly annoying feature of thermistor-stabilized circuits is that, in variable-frequency applications, the output amplitude of the sine wave tends to judder or 'bounce' as the frequency control pot is swept up and down its range.

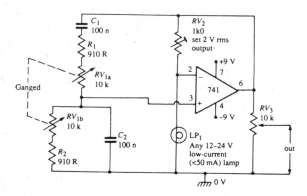

Figure 2.5   *150 Hz–1.5 kHz lamp-stabilized Wien bridge oscillator*

Diode stabilization

The amplitude 'bounce' problem of variable-frequency circuits can be minimized by using the basic circuits of *Figures 2.6 or 2.7*, which rely on the onset of diode or zener conduction for automatic gain control. In essence, RV_2 is set so that the circuit gain is slightly greater than unity when the output is close to zero, causing the circuit to oscillate, but as each half-cycle nears the desired peak value one or other of the diodes starts to conduct and thus reduces the circuit gain, automatically stabilizing the peak amplitude of the output signal. This 'limiting' technique typically results in the generation of 1% to 2% distortion on the sine-wave output. The maximum peak-to-peak output of each circuit is roughly double the breakdown voltage of its diode regulator element.

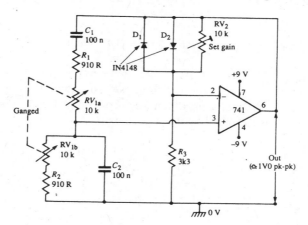

Figure 2.6 *Diode-regulated 150 Hz–1.5 kHz Wien bridge oscillator*

In the *Figure 2.6* circuit, the diodes start to conduct at about 500 mV, so the circuit gives a peak-to-peak output of about 1V0. In the *Figure 2.7* circuit, zener diodes ZD_1 and ZD_2 are connected back-to-back, and may have values as high as 5V6, giving a peak-to-peak output of about 12 V. Each circuit is set up by adjusting RV_2 to the maximum value (minimum distortion) at which oscillation is maintained across the whole frequency band.

The frequency ranges of the above circuits can be changed via the C_1 and C_2 values; increasing them by a decade reduces the frequency values by a decade, etc. *Figure 2.8* shows a variable-frequency Wien oscillator that covers the range 15 Hz to 15 kHz in three switched

20 Timer/Generator Circuits Manual

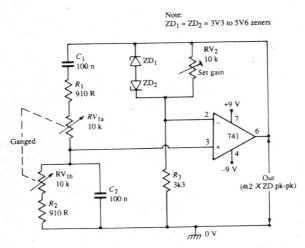

Figure 2.7 Zener-regulated 150 Hz–1.5 kHz Wien bridge oscillator

decade ranges. The circuit uses zener diode amplitude stabilization, and its output is adjustable via both switched and fully variable attenuators. Note that the maximum useful operating frequency of this type of circuit is restricted by the slew rate limitations of the op-amp;

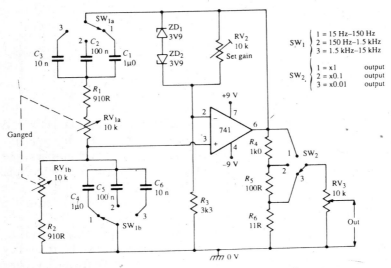

Figure 2.8 Three-decade 15 Hz–15 kHz Wien bridge oscillator

the limit is about 25 kHz with a 741 op-amp, or about 70 kHz with a CA3140.

Twin-T oscillators

Another way of making a sine-wave oscillator is to wire a twin-T network between the output and input of an inverting op-amp, as shown in *Figure 2.9*. The twin-T network comprises R_1–R_2–C_3 and C_1–C_2–R_3–RV_1, and in a balanced circuit these components are in the ratios $R_1 = R_2 = 2(R_3 + RV_1)$, and $C_1 = C_2 = C_3/2$. When the network is perfectly balanced it acts as a frequency-dependent attenuator that gives zero output at a centre frequency, f_o, of $1/2\pi.R_1.C_1$, and a finite output at all other frequencies. When the network is imperfectly balanced it gives a minimal but finite output at f_o, and the phase of this output depends in the direction of the imbalance. If the imbalance is caused by $(R_3 + RV_1)$ being too low in value, the output phase is inverted relative to the input.

In *Figure 2.9*, the twin-T network is wired between the output and the inverting input of the op-amp, and RV_1 is critically adjusted so that the twin-T gives a small phase-inverted output at the f_o of 1 kHz. Zero overall phase inversion thus occurs around the feedback loop, and the circuit oscillates at a centre frequency of 1 kHz. In practice, RV_1 is adjusted so that oscillation is barely sustained, and under this condition the sine-wave output has less than 1% distortion. Automatic

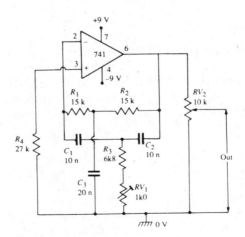

Figure 2.9   *1 kHz twin-T oscillator*

amplitude control occurs because of the progressive non-linearity of the op-amp as the output signal approaches clipping level. The output amplitude is fully variable from zero to about 5 V r.m.s via RV_2.

Figure 2.10 shows a method of amplitude control that gives slightly less distortion. Here, D_1 provides a feedback signal via potential divider RV_2. This diode progressively conducts and reduces the circuit gain when the diode forward voltage exceeds 500 mV. To set up the circuit, first set RV_2 slider to the op-amp output and adjust RV_1 so that oscillation is just sustained; under this condition the output signal has an amplitude of about 500 mV peak-to-peak. RV_2 then enables the output signal to be varied between 170 mV and 3 V r.m.s.

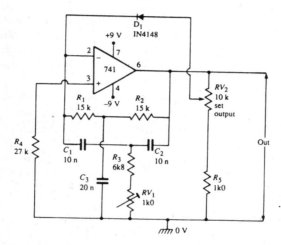

Figure 2.10 *Diode-regulated 1 kHz twin-T oscillator*

Note that these twin-T circuits make good fixed-frequency oscillators, but are not recommended for variable-frequency use, due to the difficulties of simultaneously varying three or four network components.

L–C oscillator circuits

C–R sine-wave oscillators are useful for generating signals ranging from a few hertz up to several tens or hundreds of kilohertz. L–C oscillators, on the other hand, are useful for generating signals ranging from a few tens of kilohertz to hundreds of megahertz. *Figures 2.11 to*

2.16 show a selection of practical transistor-based L–C oscillator circuits.

A transistor L–C oscillator consists, in essence, of a simple transistor amplifier stage plus a frequency-selective L–C network that gives appropriate positive feedback between its output and input. L–C networks have inherently high 'Q' or frequency-selectivity, so such oscillators consequently tend to produce reasonably pure sine-wave outputs, even when the oscillator's loop gain is far greater than unity.

Many different versions of the transistor L–C oscillator are in common use; the simplest is the **tuned collector feedback** oscillator, and an example of this is shown in *Figure 2.11*. Here, Q_1 is wired as a common emitter amplifier, with base bias provided via R_1–R_2 and with emitter resistor R_3 decoupled to high-frequency signals via C_2. L_1–C_1 form the tuned collector circuit, and collector-to-base feedback is provided via L_2, a small winding inductively coupled to L_1 and which thus provides a transformer action. By selecting the phase of this feedback signal, the circuit can be made to give zero loop phase shift at the tuned

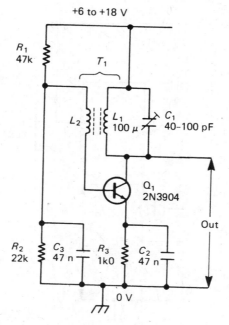

Figure 2.11 *Tuned-collector feedback L–C oscillator*

frequency so that, if the loop gain (determined by T_1's turns ratio) is greater than unity, the circuit will oscillate.

A feature of any *L–C* tuned circuit is that the phase relationship between its energizing current and induced voltage varies over the range $-90°$ to $+90°$, and equals zero at a 'centre' frequency given by $f = 1/(2\pi\sqrt{LC})$. The *Figure 2.11* circuit gives zero overall phase shift and thus oscillates at this centre frequency; with the component values shown the frequency can be varied from 1 MHz to 2 MHz via C_1, but in practice the basic circuit can easily be designed to operate at frequencies ranging from tens of hertz (by using a laminated iron-core transformer) up to tens or hundreds of megahertz.

Circuit variations

Figure 2.12 shows a simple variation of the *Figure 2.11* design, this particular circuit being known as a Hartley oscillator. Here, collector load inductor L_1 is tapped 20% down from its top, and the circuit's

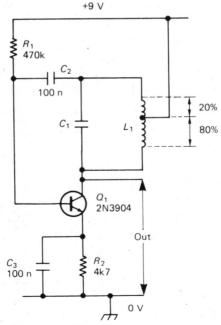

Figure 2.12 *Basic Hartley L–C oscillator*

positive supply rail is connected to this tap point; L_1 thus gives an autotransformer action in which the signal voltage appearing at the top of L_1 is 180° out of phase with that on its low (Q_1 collector) end. The signal voltage from the top of the coil (which is 180° out of phase with the collector signal) is coupled to Q_1 base via isolating capacitor C_2, and the circuit thus oscillates at a centre frequency determined by the L–C values.

Note from the above description that oscillator action depends on a *common signal* tapping point being made into the tuned circuit, so that a phase-splitting autotransformer action is obtained. This tapping point does not in fact have to be made in the actual tuning coil, but can be made into the tuning capacitor, as in the Colpitts oscillator circuit shown in *Figure 2.13*. With the component values shown this particular circuit oscillates at about 37 kHz.

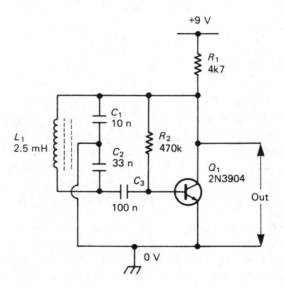

Figure 2.13   *37 kHz Colpitts L–C oscillator*

Note in *Figure 2.13* that C_1 is in parallel with Q_1's output capacitance, and C_2 is in parallel with Q_1's input capacitance. Consequently, changes in Q_1 capacitance (due to thermal shifts, etc.) cause a change in frequency. This effect can be minimized (and good frequency stability obtained) by making C_1 and C_2 large relative to the internal capacitances of Q_1.

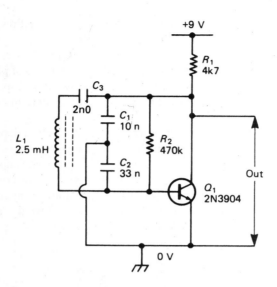

Figure 2.14 *80 kHz Gouriet or Clapp L–C oscillator*

A modification of the **Colpitts oscillator**, known as the **Clapp or Gouriet oscillator**, is shown in *Figure 2.14*. Here, a further capacitor (C_3) is wired in series with L_1, and has a value that is small relative to C_1 and C_2. Consequently, the circuit's resonant frequency is determined mainly by the values of L_1 and C_3, and is almost independent of variations in transistor capacitances. This circuit thus gives excellent frequency stability. With the component values shown, it oscillates at about 80 kHz.

Figure 2.15 shows the basic circuit of a so-called **Reinartz oscillator**. Here, the tuning coil has three inductively coupled windings. Positive feedback is obtained by coupling the collector and emitter signals of the transistor via windings L_1 and L_2. Both of these inductors are coupled to L_3, and the circuit oscillates at a frequency determined by L_3–C_1. The diagram shows typical coil-turns ratios for a circuit designed to oscillate at a few hundred kilohertz.

Waveform modulation

To complete this look at basic *L–C* oscillators, *Figures 2.16* and *2.17* show how the *Figure 2.11* design can be modified so that it acts as a

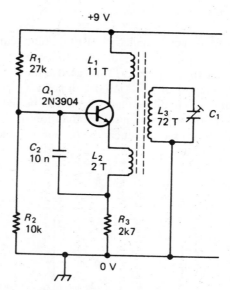

Figure 2.15 *Basic Reinartz L–C oscillator*

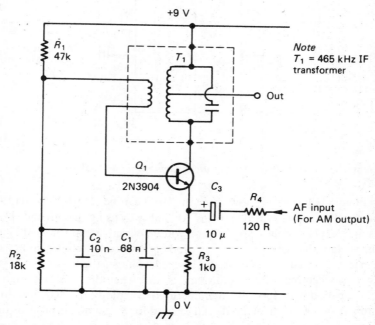

Figure 2.16 *465 kHz BFO with amplitude-modulation facility*

465 kHz beat-frequency oscillator (BFO) that can produce a modulated (either AM or FM) output. In both cases a standard 465 kHz transistor IF transformer (T_1) is used as the L–C tuned circuit, and emitter biasing resistor R_3 is RF decoupled via C_1.

In the *Figure 2.16* circuit an audio signal can be fed to Q_1 emitter via R_4 and C_3, to provide amplitude modulation of the 465 kHz RF output signal. This simple circuit can be used to produce modulation depths up to about 40%.

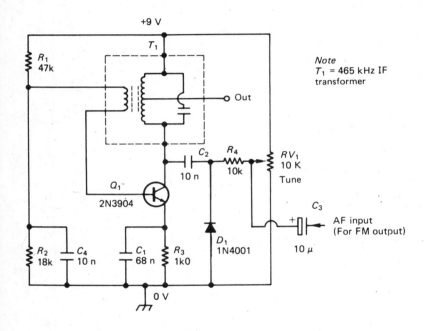

Figure 2.17 *465 kHz BFO with varicap tuning and FM facility*

Finally, *Figure 2.17* shows how the BFO can be tuned via potentiometer RV_1 and provided with an FM facility by using silicon diode D_1 as an inexpensive varicap diode or voltage-variable capacitor. It is a simple fact that when any silicon diode is reverse biased it exhibits a capacitance that varies with the applied voltage; the capacitance is greatest when the voltage is low, and is least when the voltage is high. Varicap diodes are specially manufactured to exploit this effect, but the ordinary IN4001 silicon diode can be used for the same purpose, as in *Figure 2.17*.

Here, C_2 (which gives d.c. isolation between Q_1 and D_1) and *capacitor* D_1 are wired in series, and the combination is effectively wired across the T_1 tuned circuit (since the circuit's supply rails are shorted together as far as ac signals are concerned). Consequently, the oscillator's centre frequency can be varied by simply altering the capacitance of D_1 via RV_1. Similarly, the oscillator frequency can be modulated around its centre value by feeding an external audio signal to the R_4–RV_1 junction via C_3, as shown, to provide the FM facility.

3 Square-wave generators

Square waves can be regarded as free-running pulse waveforms and are very easy to produce, either by conversion from existing waveforms or by direct production from various types of generator circuit. Fifty such circuits are described in this chapter.

Sine-to-square converters

A good square wave can be generated by feeding an existing sine wave through a sine-to-square converting Schmitt trigger. *Figure 3.1* shows a transistor circuit of this kind; it needs a sine-wave input of 0.5 V r.m.s. or greater, produces square-wave outputs with rise times of about 250 ns when the output is lightly loaded, and has a good performance up to several hundred kilohertz. In use, RV_1 should be adjusted to give the best output waveform symmetry.

Figure 3.2 shows a CMOS version of the sine-to-square converter. It uses only one of the four available gates of a 4093B quad 2-input NAND Schmitt IC, the other three gates being disabled by grounding their inputs. This circuit produces an excellent square-wave output, with typical rise and fall times of less than 100 ns when the output is loaded by 50 pF.

Transistor astable circuits

One way of directly generating square waves is via a two-transistor astable multivibrator, which can operate from supply values as low as 1.5 V or (with slight modification) as high as several tens of volts. *Figure*

Square-wave generators 31

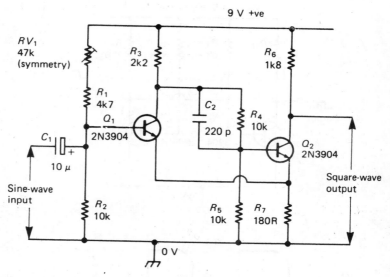

Figure 3.1   *Transistor Schmitt sine/**square** converter*

3.3 shows a practical 1 kHz version of this circuit, together with its basic waveforms.

The *Figure 3.3* circuit acts as a self-oscillating regenerative switch, in which the on and off periods are controlled by the C_1–R_1 and C_2–R_2 time constants. If these time constants are equal ($C_1 = C_2$ and $R_1 = R_2$),

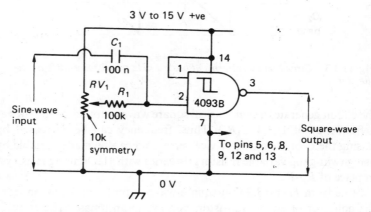

Figure 3.2   *CMOS Schmitt sine/square converter*

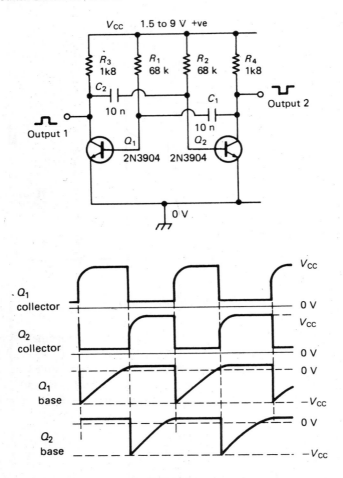

Figure 3.3 *Circuit and waveforms of basic 1 kHz transistor astable multivibrator*

the circuit generates a symmetrical square wave output and operates at a frequency of $1/(1.4\ C_1 R_1)$. Thus, frequency can be decreased by raising the C or R values, or vice versa, and can be made variable by using twin-gang variable resistors (in series with 10k limiting resistors) in place of R_1–R_2.

Note from *Figure 3.3* that square-wave outputs can be taken from the collector of either transistor, and are in antiphase. The circuit's operating frequency is almost independent of supply line values in the range 1.5 to 9 V. The upper voltage limit is set by the fact that, as the transistors switch regeneratively at the end of each half-cycle, the base-emitter junction of the off-going transistor is reverse biased by an amount equal to the supply line value, and if this exceeds the transistor's reverse base-emitter breakdown value (usually about 9 V), the timing operations are upset. This snag can be overcome by using the modifications shown in *Figure 3.4*.

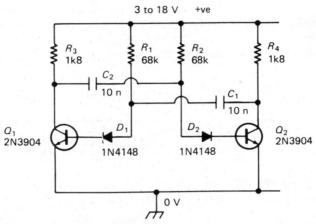

Figure 3.4 *1 kHz astable with frequency-correcting protection diodes on the transistor inputs*

Here, a 1N4148 silicon diode is wired in series with the base input of each transistor and effectively raises its reverse base-emitter breakdown voltage to about 80 V. Consequently, the maximum supply value of this circuit is limited only by the collector-emitter breakdown value of the transistors, and may be several tens of volts. In practice, the 'protected' circuit of *Figure 3.4* gives a frequency variation of only 2% when the supply value is varied from 6 to 18 V.

Note that the output waveform leading edges of these two circuits are rounded and have rather long rise times, so these simple astables are in fact only useful as 'squirt' generators. In practice, the lower the values of R_1 and R_2 relative to those of R_3 and R_4, the worse this rounding becomes, so R_1 and R_2 should be as large as possible, but must be limited to maximum values equal to the products of transistor current gain (say, 90) and the R_3 (or R_4) values (1.8k in this case); thus, the maximum possible values of R_1 and R_2 are 162k in the *Figure 3.3* and *3.4* circuits.

Circuit modifications

The rounding of the leading edges of the basic astable are caused by the loading effects of each cross-coupled timing capacitor, which stops the collector voltage from switching rapidly to the positive rail as its transistor turns off. This action can be overcome, and excellent *clock quality* square waves obtained, by using diodes to effectively isolate each capacitor from the collector of its transistor as it turns off, as in the 1 kHz generator of *Figure 3.5*. Here, D_1 and D_2 are used to isolate the timing capacitors at the moment of regenerative switching. The circuit's main time constants are set by C_1–R_1 and C_2–R_2, and the effective collector loads of Q_1 and Q_2 are equal to the parallel resistances of R_3–R_4 and R_5–R_6 respectively.

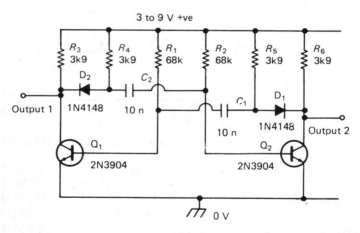

Figure 3.5 *1 kHz astable with waveform correction applied via D_1 and D_2*

The basic operation of the astable multivibrator depends on slight imbalances between the transistor characteristics, so that one transistor turns on slightly quicker than the other when power is first applied. If the circuit's supply is applied by slowly raising it from zero, both transistors may turn on simultaneously, in which case oscillation will not occur. This snag can be overcome by using the sure-start circuit of *Figure 3.6*, in which the timing resistors are connected to the transistor collectors in such a way that only one transistor can ever be turned on at any given moment.

The transistor astable circuits shown so far are designed to give a

Square-wave generators

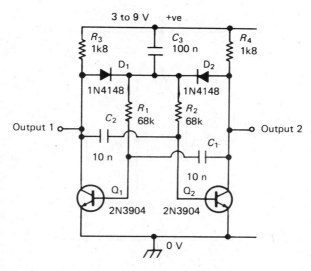

Figure 3.6 *1 kHz astable with sure-start facility*

symmetrical output waveform, with a 1:1 M/S ratio. A non-symmetrical waveform can be obtained by making one set of astable time-constant components larger than the other. *Figure 3.7* shows the connections for making a fixed-frequency (about 1100 Hz) variable M/S ratio generator, in which the ratio is fully variable from 1:10 to 10:1.

In some applications the leading edges of the waveforms of the above circuit may be considered excessively rounded when the mark/space control is set to its extreme positions. Also, the circuit may refuse to start if its supply voltage is applied too slowly. Both of these snags can be overcome by using the circuit of *Figure 3.8*, which is fitted with both sure-start and waveform correction diodes.

Op-amp square-wave generators

Another way of generating square waves is to use an op-amp wired in the basic relaxation oscillator configuration shown in *Figure 3.9*. This circuit uses dual power supplies, and its output switches alternately between the positive and negative saturation levels of the op-amp. Potential divider R_2–R_3 feeds a fraction of this voltage back to the non-inverting input of the op-amp, to provide the circuit with an

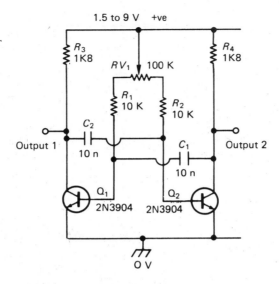

Figure 3.7 *Astable with variable M/S ratio*

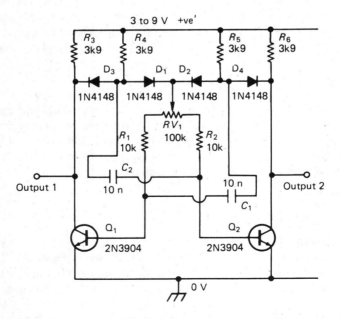

Figure 3.8 *Improved variable M/S ratio astable with waveform correction and sure-start facility*

Square-wave generators   37

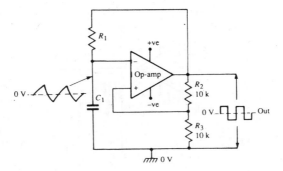

Figure 3.9   *Basic op-amp relaxation oscillator circuit*

'aiming' voltage, and feedback components R_1–C_1 act as a time-constant network.

The basic operation of *Figure 3.9* is such that, when the output is high, C_1 charges up via R_1 until its voltage reaches the positive 'aiming' value set by R_2–R_3, at which point a comparator action occurs and the op-amp output regeneratively switches negative, causing C_1 to start to discharge via R_1 until its voltage falls to the negative aiming value set by R_2–R_3, at which point the op-amp output switches positive again, and the whole sequence repeats *ad infinitum*. The action is such that a symmetrical square wave is developed at the output of the op-amp, and a non-linear triangle waveform is developed across C_1. These waveforms swing symmetrically either side of the zero-volts line. A fast op-amp, such as the CA3140, should be used if good rise and fall times are needed from the square wave.

This circuit's operating frequency can be varied by altering either the R_1 or C_1 values, or by altering the R_2–R_3 ratios; the circuit is thus quite versatile. *Figure 3.10* shows how it can be adapted to make a 500 Hz to 5 kHz square-wave generator, with frequency variation obtained by altering the attenuation ratio of R_2–RV_1–R_3. *Figure 3.11* shows how the circuit can be improved by using RV_2 to pre-set the range of the RV_1 frequency control, and by using RV_3 as an output amplitude control.

Figure 3.12 shows the above circuit modified to make a general-purpose square-wave generator that spans 2 Hz to 20 kHz in four switched decade ranges. Pre-set pots RV_1 to RV_4 are used to precisely set the minimum frequency of the 2 Hz–20 Hz, 20 Hz–200 Hz, 200 Hz–2 kHz, and 2 kHz–20 kHz ranges respectively.

38 Timer/Generator Circuits Manual

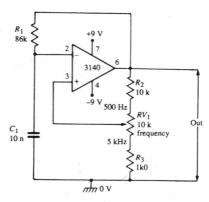

Figure 3.10 *Simple 500 Hz–5 kHz op-amp square-wave generator*

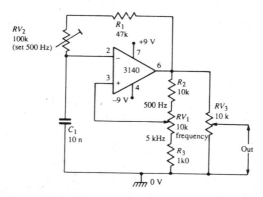

Figure 3.11 *Improved 500 Hz–5 kHz square-wave generator*

Variable symmetry

In the *Figure 3.9* circuit, C_1 alternately charges and discharges via R_1, and the circuit generates a symmetrical square-wave output. The circuit can easily be modified to give a variable-symmetry output by providing C_1 with alternate charge and discharge paths, as shown in *Figures 3.13* and *3.14*.

In *Figure 3.13* the waveform's M/S ratio is fully variable from 11:1 to 1:11 via RV_1, and the frequency is variable from 650 Hz to 6.5 kHz via RV_2. The action is such that C_1 alternately charges up via R_1–D_1 and the left-hand side of RV_1, and discharges via R_1–D_2 and the right-hand

Square-wave generators 39

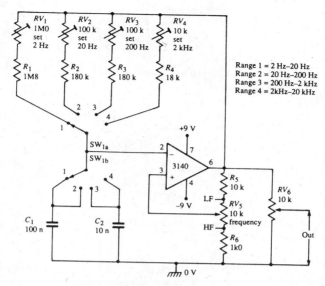

Figure 3.12 *General-purpose four-decade 2 Hz–20 kHz op-amp square-wave generator*

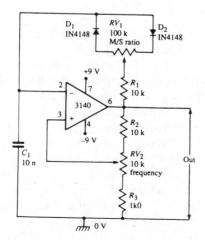

Figure 3.13 *Square-wave generator with variable M/S ratio and frequency*

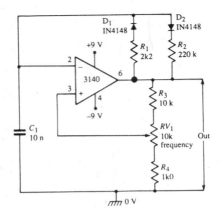

Figure 3.14 *Variable-frequency narrow pulse generator*

side of RV_1, giving a variable-symmetry output; variation of RV_1 has negligible effect on the circuit's operating frequency.

In *Figure 3.14* the mark period is determined by C_1–D_1–R_1 and the space period by C_1–D_2–R_2; these periods differ by a factor of one hundred, so the circuit generates a narrow pulse waveform; pulse frequency is variable from 300 Hz to 3 kHz via RV_1.

Resistance activation

Note from the description of the *Figure 3.9* oscillator that the circuit actually changes state in each half-cycle at the point where the C_1 voltage reaches the 'aiming' value set by the R_2–R_3 potential divider; if C_1 is unable to attain this voltage, the circuit will not oscillate. Thus, if the circuit is modified as shown in *Figure 3.15*, in which RV_1 is wired in parallel with C_1 and forms a potential divider with R_1, and R_2–R_3 have a 1:1 ratio, the circuit will oscillate only if RV_1 has a value greater than R_1. This circuit can thus function as a **resistance-activated oscillator**.

Figures 3.16 and *3.17* show two practical applications of the resistance-activated oscillator. The *Figure 3.16* circuit acts as a precision light-activated oscillator or alarm, and uses an LDR as the resistance-activating element; the circuit can be converted to a *dark-activated* oscillator by transposing by LDR–RV_1 positions. The *Figure 3.17* circuit uses NTC thermistor TH_1 as the resistance-activating element, and acts as a precision *over-temperature* oscillator/alarm; the

Square-wave generators 41

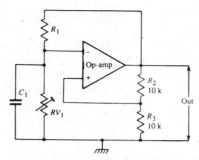

Figure 3.15 *Basic resistance-activated relaxation oscillator*

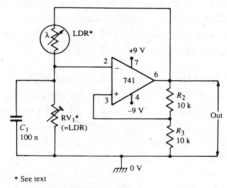

* See text

Figure 3.16 *Precision light-activated oscillator/alarm*

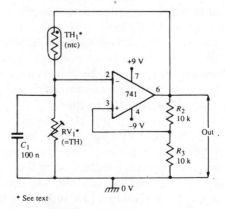

* See text

Figure 3.17 *Precision over-temperature oscillator/alarm*

circuit can be converted to an *under-temperature* oscillator by transposing TH_1 and RV_1.

In the above two circuits the LDR or TH_1 can have any resistance in the range 2k0 to 2M0 at the required trigger level, and RV_1 must have the same value as the activating element at the desired trigger level. RV_1 sets the trigger level; the C_1 value can be altered to change the oscillator frequency.

Op-amp circuit variations

The basic *Figure 3.9* op-amp relaxation oscillator circuit is shown for use with dual power supplies. *Figure 3.18* shows how the basic design can be modified for use with single-ended supplies, to make a square-wave generator that gives an output that switches between 0 V and roughly 2 V below the positive supply rail value. This circuit operates as follows.

Suppose that the output has just switched high; in this case R_3 is effectively switched in parallel with R_1, so roughly two-thirds of the supply voltage is applied to pin 3, and C_1 charges towards the positive supply rail via R_4 and the op-amp output until the C_1 voltage reaches this pin 3 value, at which point a regenerative switching action is initiated and causes the op-amp output to switch abruptly to 0 V.

Under this new condition R_3 is effectively switched in parallel with R_2, so only one-third of the supply voltage is applied to pin 3, and C_1 starts to discharge towards zero via R_4 and the op-amp output until C_1 voltage reaches the new pin 3 value and another regenerative switching action is initiated in which the output switches abruptly high again. The process then repeats *ad infinitum*.

The circuit's waveform period is determined by the R_3, R_4 and C_1 values, and is about 6 ms with the component values shown. The period can be increased (or reduced) by increasing (or reducing) the values of C_1 and/or R_4; C_1 can have any value from 33 pF to 1000 μF, and R_4 can have any value from 10k to 10M.

The *Figure 3.18* circuit is quite useful, but has several imperfections. Since its output does not switch to the full positive supply rail voltage when in the 'high' state, its output waveform is not quite symmetrical, and its period and symmetry vary slightly when the supply voltage is varied. Also, the waveform rise and fall times are limited by the slew rate of the op-amp and, when using a 15 V supply, have typical values of about 12 μs and 7 μs respectively when feeding a 50 pF load.

Square-wave generators 43

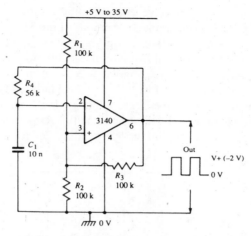

Figure 3.18 *Op-amp relaxation oscillator modified for operation from a single-ended supply*

Figure 3.19 shows an improved version of the circuit, in which the op-amp and R_5–Q_1–R_6 act together to form a high-performance compound op-amp. This circuit is free of the defects mentioned above; its output waveform is quite symmetrical, switches fully between the zero and positive supply rails, has a period that is independent of supply rail values, and has typical rise and fall times of 1 μs and 0.7 μs. The circuit

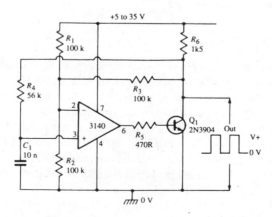

Figure 3.19 *High-performance compound op-amp square-wave generator circuit*

44 Timer/Generator Circuits Manual

is similar to that described above, except for the addition of Q_1–R_5–R_6 and the transposing of the input connections of the 3140 op-amp.

Finally, to complete this look at op-amp square-wave generators, *Figure 3.20* shows how the above circuit can be modified so that it generates a fixed-frequency rectangular waveform in which the M/S ratio is fully variable from 25:1 to 1:25 via RV_1. The circuit operation is similar to that already described, except that on 'high' parts of the cycle C_1 charges via R_4–D_2 and the right-hand half of RV_1, and on the 'low' parts discharges via R_4–D_1 and the left-hand half of RV_1.

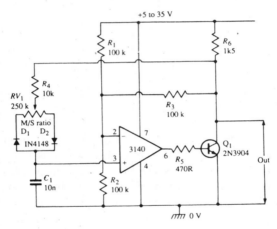

Figure 3.20 *High-performance square-wave generator with M/S ratio variable from 1:25 to 25:1 via RV_1*

CMOS astable basics

Another way to make a good square-wave generator is to use the gates of inexpensive CMOS logic ICs such as the 4001B, 4011B, or 4093B, as simple inverter stages, and to then wire those inverters in the astable multivibrator mode, as shown in the simple astable circuit of *Figure 3.21(a)*. This circuit generates a good square-wave output from IC_{1b} (and a not-quite-so-good antiphase square-wave output from IC_{1a}), and operates at about 1 kHz with the component values shown. The circuit is suitable for use in many (but not all) 'clock' generator applications, and operates as follows.

In *Figure 3.21(a)*, the two inverters are wired in series, so the output of one goes high when the other goes low, and vice versa. Time-constant network C_1–R_1 is wired between the outputs of IC_{1b} and IC_{1a},

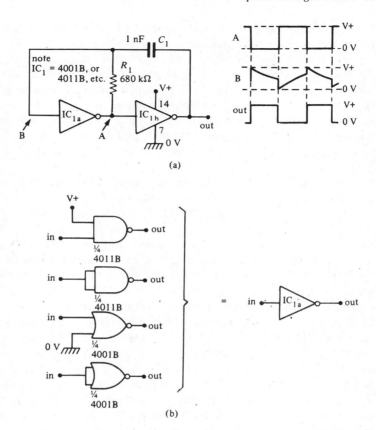

Figure 3.21 (a) *Circuit and waveforms of basic two-stage 1 kHz CMOS astable.* (b) *Ways of connecting a two-input NAND (4011B) or NOR (4001B) gate for use as an inverter*

with the C_1–R_1 junction fed to the input of the IC_{1a} inverter stage. Suppose initially that C_1 is fully discharged, and that the output of IC_{1b} has just switched high (and the output of IC_{1a} has just switched low).

Under this condition the C_1–R_1 junction is initially at full positive supply volts, thus driving IC_{1a} output low, but starts to decay exponentially as C_1 charges up via R_1, until eventually it falls into the linear transfer voltage range of IC_{1a}, making its output start to swing high. This swing is amplified by inverter IC_{1b}, initiating a regenerative action in which IC_{1b} output switches abruptly to the low state (and IC_{1a} output switches high). This switching action makes the charge of C_1 try to apply a negative voltage of IC_{1a} input, but IC_{1a}'s built-in input protection diodes prevent this and instead discharge C_1.

Thus, at the start of the second cycle, C_1 is again fully discharged, so the C_1–R_1 junction is initially at zero volts (driving IC_{1a} output high), but then rises exponentially as C_1 charges up via R_1, until eventually it rises into the linear 'transfer voltage' range of IC_{1a}, thus initiating another regenerative switching action in which IC_{1b} output switches high again (and IC_{1a} output switches low), and C_1 is again discharged via the IC_{1a} input protection diodes. The operating cycle then continues *ad infinitum*.

The circuit's operating frequency is inversely proportional to the C–R time constant (the period is roughly $1.4 \times C.R$), so can be raised by lowering the values of either C_1 or R_1. C_1 must be non-polarized and can vary from a few tens of picofarads to several microfarads, and R_1 can vary from 4k7 to 22M; the astable operating frequency can vary from a fraction of a hertz to about 1 MHz. For variable-frequency operation, wire a fixed and a variable resistor in series in the R_1 position..

Note at this point that each of the 'inverters' of the *Figure 3.21(a)* circuit can be made from a single gate of a 4001B quad 2-input NOR gate or a 4011B quad 2-input NAND gate etc., by using the connections shown in *Figure 3.21(b)*. Thus each of these ICs can provide two astable circuits. Also note that the inputs of all unused gates in these ICs must be tied to one or other of the supply-line terminals. The *Figure 3.21(a)* astable (and all other CMOS astables shown in this chapter) can be used with any supplies in the range 3 V to 18 V; the zero volts terminal goes to pin 7 of the 4001B or 4011B, and the positive terminal goes to pin 14.

The output of the *Figure 3.21(a)* astable switches (when lightly loaded) almost fully between the zero and positive supply rail values, but the C_1–R_1 junction voltage is prevented from swinging below zero or above the positive supply-rail levels by the built-in clamping diodes at the input of IC_{1a}. This factor makes the operating frequency somewhat dependent on supply rail voltages. Typically, the frequency falls by about 0.8% for a 10% rise in supply voltage. If the frequency is normalized with a 10 V supply, the frequency falls by 4% at 15 V or rises by 8% at 5 V.

The operating frequency of the *Figure 3.21(a)* circuit is also influenced by the 'transfer voltage' value of the individual IC_{1a} inverter/gate that is used in the astable, and may vary by as much as 10% between different ICs. The square-wave's output symmetry also depends on the transfer voltage value, and in most cases the circuit will give a non-symmetrical output. In most non-precision and hobby applications these defects are, however, of little practical importance.

Astable variations

Some of the defects of the *Figure 3.21(a)* circuit can be minimized by using the compensated astable of *Figure 3.22*, in which R_2 is wired in series with IC_{1a}'s input. This resistor must be large relative to R_1, and its main purpose is to allow the C_1–R_1 junction to swing freely below the zero and above the positive supply rail voltages and thus improve the astable's frequency stability. Typically, when R_2 is ten times the value of R_1, the frequency varies by only 0.5% when the supply voltage is varied between 5 and 15 V. An incidental benefit of R_2 is that it gives a slight improvement in waveform symmetry.

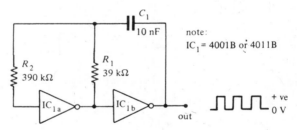

Figure 3.22 *This compensated version of the 1 kHz astable has excellent frequency stability*

The basic and compensated astable circuits of *Figures 3.21(a)* and *3.22* can be built with several detail variations, as shown in *Figures 3.23* to *3.26*. In the basic astable circuit, for example, C_1 alternately charges and discharges via R_1 and thus generates a fixed symmetry output. *Figures 3.23* to *3.25* show how the basic circuit can be modified to give alternate C_1 charge and discharge paths and to thus allow the symmetry to be varied at will.

The *Figure 3.23* circuit generates a highly non-symmetrical waveform, equivalent to a fixed pulse delivered at a fixed *time base rate*. Here, C_1 charges in one direction via R_2 in parallel with the D_1–R_1 combination, to generate the mark part of the waveform, but discharges in the reverse direction via R_2 only, to give the space part of the waveform.

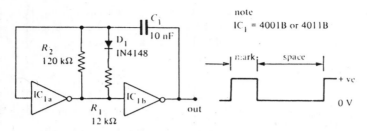

Figure 3.23 Modified CMOS astable with non-symmetrical output

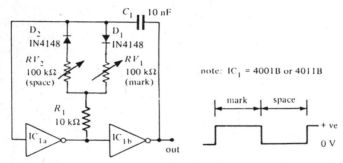

Figure 3.24 CMOS astable with independently variable mark and space times

Figure 3.24 shows the modifications for generating a waveform with independently-variable mark and space times; the mark is controlled by R_1–RV_1–D_1, and the space by R_1–RV_2–D_2.

Figure 3.25 shows the modifications to give a variable symmetry or M/S ratio output while maintaining a near-constant frequency. Here, C_1 charges in one direction via D_2 and the lower half of RV_1 and R_2, and in the other direction via D_1 and the upper half of RV_1 and R_1. The M/S ratio can be varied over the range 1:11 to 11:1 via RV_1.

Finally, *Figures 3.26* and *3.27* show ways of using the basic astable circuit as a very simple VCO. The *Figure 3.26* circuit can be used to vary the operating frequency over a limited range via an external voltage. R_2 must be at least twice as large as R_1 for satisfactory operation, the actual value depending on the required frequency-shift range: a 'low' R_1 value gives a large shift range, and a 'large' R_2 value gives a small one. The *Figure 3.27* circuit acts as a special-effects VCO in which the oscillator frequency rises with input voltage, but switches fully off when the input voltage falls below a value pre-set by RV_1.

Square-wave generators 49

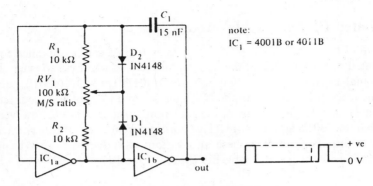

Figure 3.25 *CMOS astable with fully variable M/S ratio*

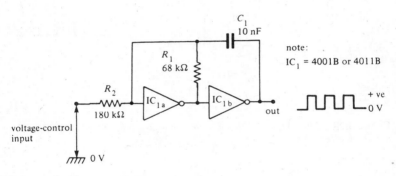

Figure 3.26 *Simple VCO circuit*

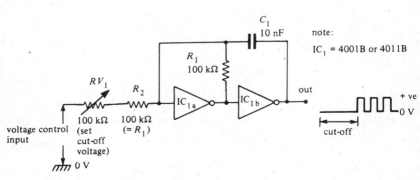

Figure 3.27 *Special-effects VCO which cuts off when V_{in} falls below a preset value*

Gated CMOS astable circuits

All of the astable circuits of *Figures 3.21* to *3.25* can be modified for gated operation, so that they can be turned on and off via an external signal, by simply using a 2-input NAND (4011B) or NOR (4001B) gate in place of the inverter in the IC_{1a} position and by applying the input gate control signal to one of the gate input terminals. Note, however, that the 4001B and the 4011B give quite different types of gate control and output operation in these applications, as shown by the two basic versions of the gated astable in *Figures 3.28* and *Figure 3.29*.

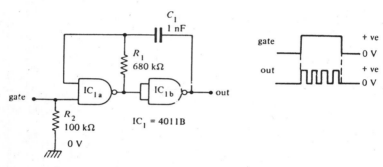

Figure 3.28 *This gated astable has a normally-low output and is gated on by a high (logic-1) input*

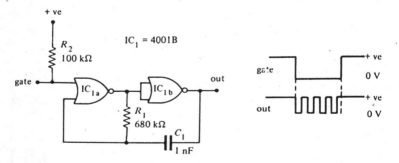

Figure 3.29 *This version of the gated astable has a normally-high output and is gated on by a low (logic-0) input*

Note that the NAND version of the circuit is gated on by a logic-1 input and has a normally-low output, while the NOR version is gated on by a logic-0 input and has a normally-high output. R_2 can be eliminated from these circuits if the gate drive is direct-coupled from the output of a preceding CMOS logic stage, etc.

Also note in these basic gated astable circuits that the output signal terminates as soon as the gate drive is removed. Thus, any noise present at the gate terminal also appears at the outputs of these circuits. *Figures 3.30* and *3.31* show how to modify the circuits so that they produce noiseless outputs.

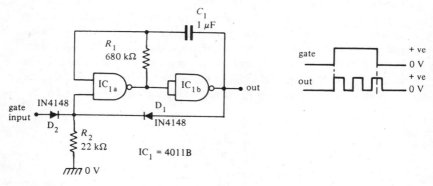

Figure 3.30 *Semi-latching or noiseless gated astable circuit, with logic-1 gate input and normally-zero output*

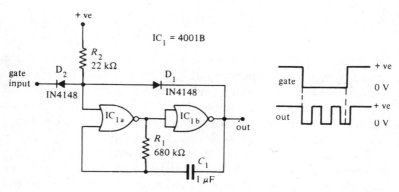

Figure 3.31 *Alternative semi-latching gated astable, with logic-0 gate input and normally-high output*

Here, IC_{1a}'s gate signal is derived from both the outside world and from IC_{1b} output via diode OR gate D_1–D_2–R_2. As soon as the circuit is gated from the outside world via D_2 the output of IC_{1b} reinforces or self-latches the gating via D_1 for the duration of one half astable cycle, thus eliminating any effects of a noisy outside world signal. The *outputs* of these *semi-latching* gated astable circuits are thus always complete numbers of half cycles.

Ring-of-three astable

The two-stage astable circuit is a good general-purpose square-wave generator, but is not always suitable for direct use as a clock generator with fast-acting counting and dividing circuits, since it tends to pick up and amplify any existing supply-line noise during the 'transitioning' parts of its operating cycle and to thus produce output square-waves with 'glitchy' leading and trailing edges. A far better type of clock generator circuit is the ring-of-three astable shown in *Figure 3.32*.

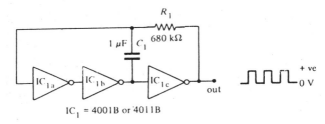

Figure 3.32 *This ring-of-three astable makes an excellent clock generator*

The *Figure 3.32* ring-of-three circuit is similar to the basic two-stage astable, except that its input stage (IC_{1a}–IC_{1b}) acts as an ultra-high-gain, non-inverting amplifier and its main timing components (C_1–R_1) are transposed (relative to the two-stage astable). Because of the very high overall gain of the circuit, it produces an excellent and glitch-free square-wave output, ideal for clock-generator use.

The basic ring-of-three astable can be subjected to all the design modifications already described for the basic two-stage astable; for example, it can be used in either basic or compensated form and can give either a symmetrical or non-symmetrical output. The most interesting variations occur, however, when the circuit is used in the gated mode, since it can be gated via either the IC_{1b} or IC_{1c} stages. *Figures 3.33* to *3.36* show four variations on this gating theme.

Thus, the *Figures 3.33* and *3.34* circuits are both gated on by a logic-1 input signal, but the *Figure 3.33* circuit has a normally low output,

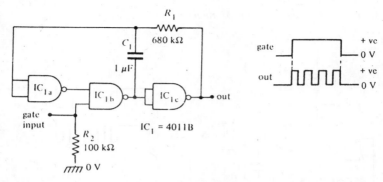

Figure 3.33 *This gated ring-of-three astable is gated by a logic-1 input and has a normally-low output*

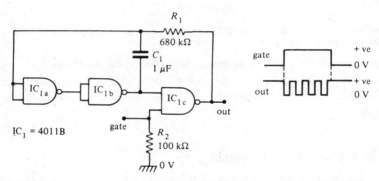

Figure 3.34 *This gated ring-of-three astable is gated by a logic-1 input and has a normally-high output*

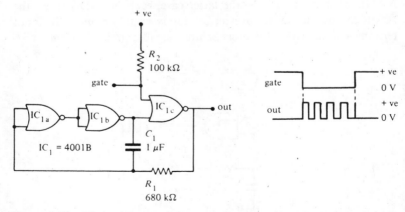

Figure 3.35 *This gated ring-of-three astable is gated by a logic-0 input and has a normally-low output*

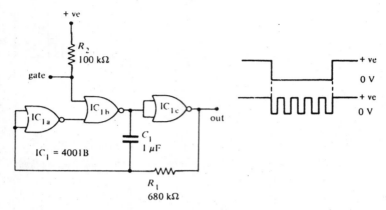

Figure 3.36 *This gated ring-of-three astable is gated by a logic-0 input and has a normally-high output*

while that of *Figure 3.34* is normally high. Similarly, the *Figure 3.35* and *3.36* circuits are both gated on by a logic-0 signal, but the output of the *Figure 3.35* circuit is normally low, while that of *Figure 3.36* is normally high.

The CMOS Schmitt astable

An excellent astable clock generator can also be made from a single CMOS Schmitt inverter stage. Suitable ICs for use in this application are the 40106B hex Schmitt inverter, and the 4093B quad 2-input NAND Schmitt trigger. In the latter case, each NAND gate of the 4093B can be used as an inverter by simply disabling one of its input terminals, as shown in the basic Schmitt astable circuit of *Figure 3.37*.

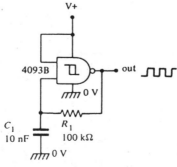

Figure 3.37 *Basic CMOS Schmitt astable*

Square-wave generators 55

The Schmitt astable gives an excellent square-wave output, with edges that are unaffected by supply line ripple and other nasties. The operating frequency is determined by the C_1–R_1 values, and can be varied from a few cycles per minute to 1 MHz or so. The circuit action is such that C_1 alternately charges and discharges via R_1, without switching the C_1 polarity; C_1 can thus be a polarized component.

Figure 3.38 shows how the 4093B-based astable can be modified so that it can be gated via an external signal; it is gated on by a high (logic-1) input, but gives a high output when it is in the gated-off state.

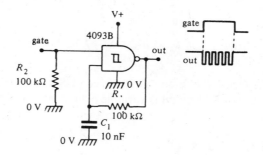

Figure 3.38 *Gated CMOS Schmitt astable*

The basic Schmitt astable circuit of *Figure 3.37* generates an inherently symmetrical square-wave output. It can be made to produce a non-symmetrical output by providing its timing capacitor with alternate charge and discharge paths, as shown in the circuits of *Figures 3.39*

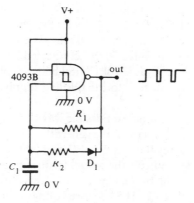

Figure 3.39 *CMOS astable with non-symmetrical M/S ratio*

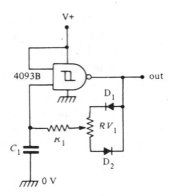

Figure 3.40 *CMOS astable with variable M/S ratio*

and *3.40*. The *Figure 3.39* circuit gives a fixed M/S ratio output; the M/S ratio of the *Figure 3.40* circuit is variable over a wide range via RV_1.

4046B VCO circuits

Before completing this look at CMOS square-wave and clock generator circuits, brief mention must be made of the 4046B PLL IC, which houses (among other things) a very useful VCO section that generates an excellent square-wave output. Full details of this IC are given in Chapter 10.

The 4046B's VCO section is highly versatile. It gives a good symmetrical square-wave output, has a top-end frequency limit in excess of 1 MHz, has a voltage-to-frequency linearity of about 1% and can be scanned through a 1000,000:1 range by an external voltage applied to the VCO input terminal. The VCO frequency is governed by the values of a capacitor (minimum value 50 pF) connected between pins 6 and 7 and a resistor (minimum value 10k) wired between pin 11 and ground, and by the voltage (from zero to the positive supply value) applied to VCO input pin 9.

Figure 3.41 shows the simplest possible way of using the 4046B VCO as a voltage-controlled square-wave generator. Here, C_1–R_1 determine the maximum frequency that can be obtained (with the pin 9 voltage at maximum) and RV_1 controls the actual frequency by applying a control voltage to pin 9: the frequency falls to a very low value (a fraction of a hertz) with pin 9 at 0 V. The effective voltage-control range of pin 9 varies from roughly 1 V below the supply value to about 1 V above

Square-wave generators 57

zero, and gives a frequency span of about 1000,000:1. Ideally, the circuit supply voltage should be regulated.

It states above that the frequency falls to near-zero when the input voltage of the *Figure 3.41* circuit is reduced to zero. *Figure 3.42* shows how the circuit can be modified so that the frequency falls all the way to zero with zero input, by wiring high-value resistor R_2 between pins 12 and 16. Note here that, when the frequency is reduced to zero, the VCO output randomly settles in either a logic-0 or a logic-1 state.

Figure 3.43 shows how the pin 12 resistor can alternatively be used to

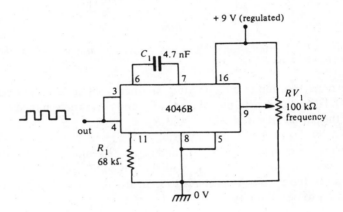

Figure 3.41 *CMOS wide-range VCO, spanning near-zero to 5 kHz via RV_1*

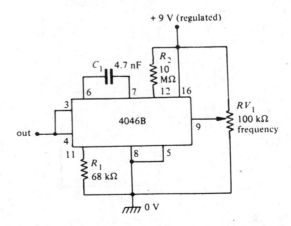

Figure 3.42 *The frequency of this VCO is variable all the way down to zero*

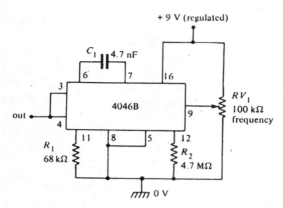

Figure 3.43 *Restricted-range VCO, with frequency variable from roughly 72 Hz to 5 kHz via RV_1*

determine the minimum operating frequency of a restricted-range VCO. Here, f_{min} is determined by C_1–R_2, and f_{max} is determined by C_1 and the parallel resistance of R_1 and R_2.

Finally, *Figure 3.44* shows a version of the restricted-range VCO in which f_{max} is controlled by C_1–R_1, and f_{min} is determined by C_1 and the series combination of R_1 and R_2; note that, by suitable choice of R_1 and R_2 values, the circuit can be made to span any desired frequency range from 1:1 to near-infinity.

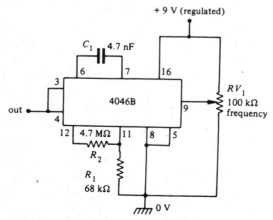

Figure 3.44 *Alternative version of the restricted-range VCO. f_{max} is controlled by C_1–R_1, f_{min} by C_1–($R_1 + R_2$)*

TTL Schmitt astable circuits

Astable square-wave generators can also be built using inexpensive TTL ICs, and one popular way of doing this is to use elements from the 74LS14 hex Schmitt inverter. *Figures 3.45* and *3.46* show examples of such circuits, which each generate a clean square-wave output with a 2:1 M/S ratio and uses a second Schmitt stage to give a buffered output. These circuits should be used with a fixed 5 V supply, and the R_1 value must be within the 100R to 1k0 range. The *Figure 3.45* design generates a fixed frequency, which has a value of about 19 kHz when C_1 equals 100 nF, and the *Figure 3.46* design generates a frequency that is

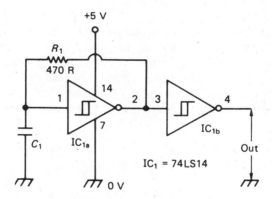

Figure 3.45 *Simple TTL Schmitt astable*

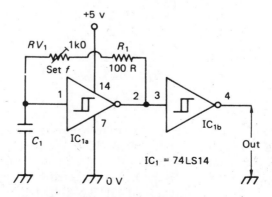

Figure 3.46 *Variable-frequency TTL Schmitt astable*

variable (via RV_1) from roughly 8.2 kHz to 89 kHz when C_1 has a value of 100 nF.

CMOS bistable circuits

Strictly speaking, a 'clock' generator can be any circuit that generates a clean (noiseless, with sharp leading and trailing edges) waveform suitable for clocking modern fast digital counter/divider circuitry, etc. Thus, a clock generator can take the form of a simple Schmitt trigger that converts a sine-wave input into a good square-wave output, or a gated or free-running square-wave generator of the type already described in this chapter, or a *monostable pulse generator* of the type described in Chapter 4.

One very useful type of clock generator is the simple R–S (reset–set) bistable (also known as the R–S flip-flop), which can be used to deliver a single clock pulse each time the available one of its two input terminals is activated, either electronically or via a push-button switch. To conclude this chapter, *Figures 3.47* to *3.50* show four ways of making such bistables, using pairs of 4001B or 4011B CMOS gates.

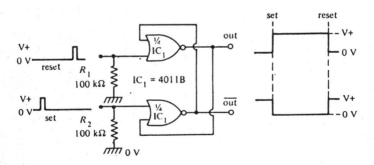

Figure 3.47 *Pulse-triggered CMOS NOR bistable*

The bistable circuit of *Figure 3.47* is triggered by positive-going input pulses, and is designed around 4001B NOR-type gates. *Figure 3.48* shows how the circuit can be modified for push-button triggering.

The bistable circuit of *Figure 3.49* is triggered by negative-going input pulses, and is designed around 4011B NAND-type gates. *Figure 3.50* shows the circuit modified for push-button triggering.

Square-wave generators 61

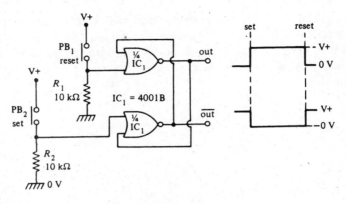

Figure 3.48 *Manually-triggered NOR bistable*

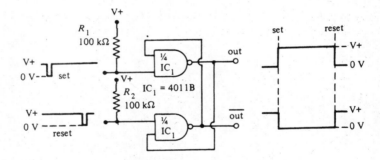

Figure 3.49 *Pulse-triggered NAND bistable*

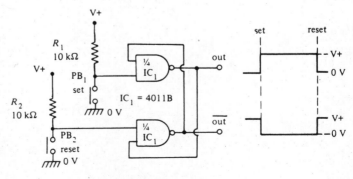

Figure 3.50 *Manually-triggered NAND bistable*

Note in the above four circuits that once the bistable has been set, its state cannot be altered until the reset terminal is activated, and once it has been reset its state can not be altered until the set terminal is activated. Also note that each circuit provides a pair of antiphase outputs.

4 Pulse generator circuits

A pulse generator is a circuit that produces a single or one-shot rectangular output waveform cycle on the arrival of an appropriate input trigger signal. Such generators may take a number of basic forms, and can be designed around a variety of semiconductor devices. In this chapter we look at the basic principles of pulse generation, and show a large selection of practical pulse generator circuits.

Pulse generator basics

The circuit designer often has to devise means of generating pulse waveforms. If the need is to simply generate a pulse of non-critical width on the arrival of the leading or trailing edge of an input square wave, a circuit element known as a **half monostable** or **edge-detector** may be used, as shown in *Figure 4.1*. Alternatively, if the need

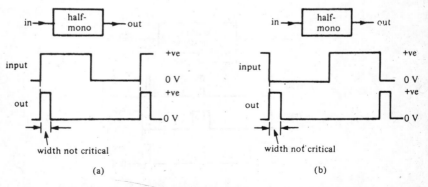

Figure 4.1 *A half-mono circuit may be used to detect (a) the leading or (b) the trailing edge of an input waveform*

is to generate a pulse of precise width on the arrival of a suitable trigger signal, a standard monostable or **one-shot multivibrator** circuit may be used.

In the standard monostable circuit, the arrival of the trigger signal initiates an internal timing cycle which causes the monostable output to change state at the start of the timing cycle, but to revert back to its original state on completion of the cycle, as shown in *Figure 4.2*.

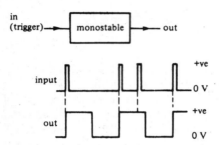

Figure 4.2 *A standard monostable generates an accurate output pulse on the arrival of a suitable trigger signal*

Note that once a timing cycle has been initiated the standard monostable circuit is immune to the effects of subsequent trigger signals until its timing period ends naturally. This type of circuit can sometimes be modified by adding a reset control terminal, as shown in *Figure 4.3*, to enable the output pulse to be terminated or aborted at any time via a suitable command signal.

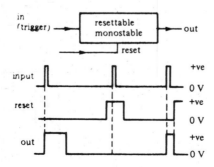

Figure 4.3 *The output pulse of a resettable mono can be aborted by a suitable reset pulse*

Pulse generator circuits

A third type of monostable circuit is the **retriggerable mono**. Here, the trigger signal actually resets the mono and then, after a very brief delay, initiates a new pulse-generating timing cycle, as shown in *Figure 4.4*, so that each new trigger signal initiates a new timing cycle, even if the trigger signal arrives in the midst of an existing cycle.

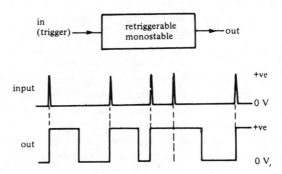

Figure 4.4 *A retriggerable mono starts a new timing cycle on the arrival of each new trigger signal*

Thus, the circuit designer may use a half mono, a standard mono, a resettable mono, or a retriggerable mono to generate pulses, the 'type' decision depending on the specific circuit design requirements.

The pulse generator may be designed around a variety of types of semiconductor device, or may be designed around a dedicated pulse generator IC; the choice is usually dictated by considerations of economics and convenience, rather than by the actual design requirements. In this chapter we look at designs based on the discrete transistor, and on readily available CMOS and TTL ICs.

Edge-detector circuits

Edge-detectors are used to generate an output pulse on the arrival of either the leading or the trailing edge of a rectangular input waveform. In most applications the precise width of the output pulse is non-critical, and in such cases the resulting circuit is known as a **half monostable** or **half mono pulse generator**.

The basic method of making an edge-detector is to feed the rectangular input waveform to a short time constant C–R differentiation network, to produce a *double-spike* output waveform (with sharp leading edges and exponential trailing edges) on the arrival of each input edge, and to then eliminate the unwanted part of the spike waveform with a discriminator diode. The remaining spike or sawtooth waveform is then converted into a clean pulse shape by feeding it through a Schmitt trigger or similar circuit. The Schmitt may be of either the inverting or the non-inverting type, depending on the required polarity of the output pulse waveform.

One of the easiest ways of making a practical edge-detector circuit is to use a CMOS Schmitt trigger IC, since these incorporate built-in protection diodes on all input terminals, and these can be used to perform the discriminator diode action described above. *Figures 4.7* to *4.10* show a selection of edge-detector designs based on CMOS Schmitt stages. Note that each gate of the popular 4093B quad 2-input NOR Schmitt IC can be used as a normal Schmitt inverter by wiring one input terminal to the positive supply rail and using the other terminal as the input point, as shown in *Figure 4.5*. A non-inverting Schmitt can be made by wiring two inverting Schmitts in series, as shown in *Figure 4.6*.

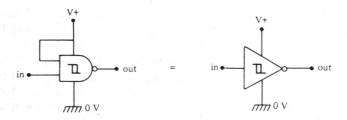

Figure 4.5 *A4093B CMOS 2-input NOR Schmitt can be used as a Schmitt inverter by wiring one input high*

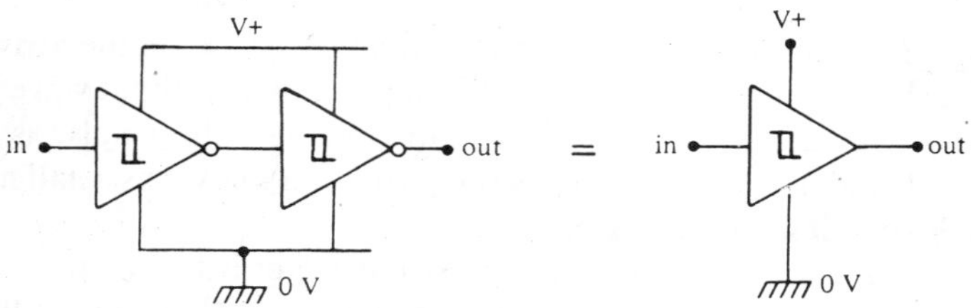

Figure 4.6 *A non-inverting Schmitt can be made by wiring two inverting Schmitt elements in series*

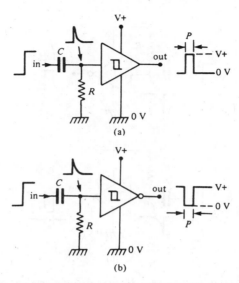

Figure 4.7 *CMOS leading-edge detector circuits giving (a) positive and (b) negative output pulses*

Figure 4.7 shows two ways of making a leading-edge detecting half-mono circuit. Here, the input of the CMOS Schmitt is tied to ground via resistor R, and $C-R$ have a time-constant that is short relative to the period of the input waveform. The leading edge of the input signal is thus converted into the 'spike' waveform shown, and this spike is then converted into a good clean pulse waveform via the Schmitt. The circuit generates a positive-going output pulse if a non-inverting CMOS Schmitt is used (*Figure 4.7(a)*), or a negative-going output pulse if an inverting Schmitt is used (*Figure 4.7(b)*). In either case, the output pulse has a period (P) of roughly $0.7\,CR$.

Figure 4.8 shows how to make a trailing-edge detecting half mono. In this case the CMOS Schmitt input is tied to the positive supply rail via R, and $C-R$ again has a short time constant. The circuit generates a positive-going output pulse if an inverting Schmitt is used (*Figure 4.8(a)*), or a negative-going pulse if a non-inverting Schmitt is used. The output pulse has a period of roughly $0.7\,CR$.

68 Timer/Generator Circuits Manual

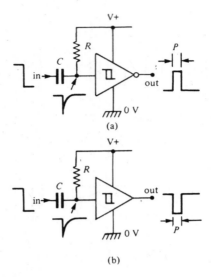

Figure 4.8 *CMOS trailing-edge detector circuit giving (a) positive and (b) negative output pulses*

Circuit variants

Two useful variants of the edge-detecting half mono circuit are the **noiseless push-button switch** of *Figure 4.9*, which effectively eliminates the adverse effects of switch contact bounce and noise, and the **power-on reset-pulse generator circuit** of *Figure 4.10*, which generates a reset pulse when power is first applied to the circuit.

In *Figure 4.9*, the input of the non-inverting CMOS Schmitt is grounded via high-value timing resistor R_1 and by input-protection resistor

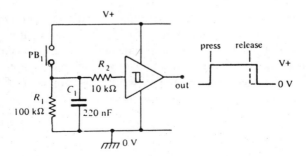

Figure 4.9 *CMOS noiseless push-button switch*

R_2, so the circuit's output is normally low. When push-button switch PB_1 is closed, C_1 charges rapidly to the full positive supply value, driving the Schmitt output high, but when PB_1 is released again C_1 discharges relatively slowly via R_1, and the Schmitt output does not return low until roughly 20 ms later. The circuit thus ignores the transient switching effects of PB_1 noise and contact bounce, etc., and generates a clean output pulse waveform with a period that is roughly 20 ms longer than the mean duration of the PB_1 switch closure.

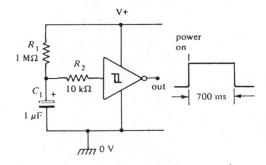

Figure 4.10   *CMOS power-on reset-pulse generator*

The *Figure 4.10* circuit uses an inverting Schmitt and produces a 700 ms output pulse (suitable for resetting external circuitry, etc.) when power is first connected. When power is initially connected C_1 is fully discharged, so the Schmitt input is pulled low and its output is switched high; C_1 then charges via R_1 until, after about 700 ms, the C_1 voltage rises to such a level that the Schmitt output switches low, completing the switch-on output pulse.

Transistor monostables

The **standard monostable** or one-shot pulse generator can be built using a variety of types of semiconductor device, and *Figure 4.11* shows the basic circuit of a manually-triggered transistor-based version of the circuit, which operates as follows.

Normally, Q_1 is driven to saturation via R_5, so the output (from Q_1 collector) is low; Q_2 (which derives its base-bias from Q_1 collector via R_3) is cut off under this condition, and its collector is at full supply-rail voltage. When a start signal is applied by briefly closing S_1, Q_1 switches

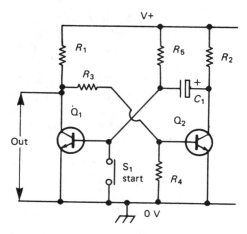

Figure 4.11 *Basic manually-triggered transistor monostable multivibrator*

off, driving the output high and driving Q_2 on via R_3, initiating a regenerative switching action in which (when S_1 reopens) Q_1 base is driven negative by the charge of C_1. As soon as the regenerative action is complete, C_1 starts to discharge via R_5 until eventually its charge falls so low that Q_1 starts to turn on again, thus initiating another regenerative action in which the transistors revert to their original states and the output pulse terminates; the action is then complete.

Thus, a positive-going output pulse is developed each time that an input trigger signal is applied via S_1. The pulse period (P) is determined by the R_5–C_1 values, and approximates $0.7 \times R_5 \times C_1$, where P is in μs, C is in microfarads, and R is in kilohms.

The basic *Figure 4.11* circuit can be triggered either manually or electronically; it can be triggered by applying a negative pulse to Q_1 base, or a positive pulse to Q_2 base. *Figure 4.12* shows the circuit and waveforms of a practical design in which manual triggering is achieved (via S_1) by feeding a positive pulse to Q_2 base via R_6.

Note in the *Figure 4.12* circuit that during part of the operating cycle the base-emitter junction of Q_1 is reverse biased by a peak amount equal to the supply voltage value, and this fact limits the maximum usable supply to about 9 V. Larger supply voltages can be used by wiring a silicon diode in series with Q_1 base, as shown, to prevent reverse base-emitter breakdown.

Also note that timing resistor R_5 must be large relative to R_2, but must be less than the product of R_1 and the current gain of Q_1. Very

Pulse generator circuits 71

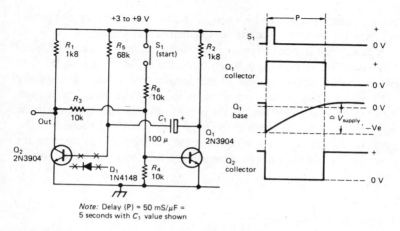

Figure 4.12 *Practical manually-triggered transistor monostable circuit*

long timing periods can be obtained by using a high-gain Darlington or super-alpha pair of transistors in place of Q_1, thus enabling large R_5 values to be used, as shown in *Figure 4.13*. This particular circuit can

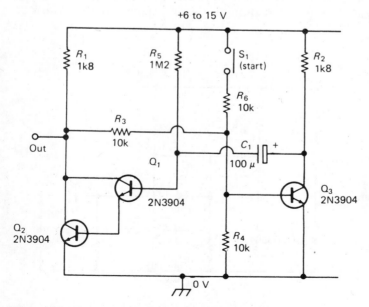

Figure 4.13 *Long-period (100 s) monostable circuit*

use any supply in the 6 to 15 V range, and gives an output pulse period of about 100 seconds with the component values shown.

A final point to note about the manually triggered circuit of *Figures 4.12* and *4.13* concerns the duration of the input trigger signal. The circuit triggers at the moment of application of a positive-going pulse to Q_2 base; if this pulse is removed by the time the monostable completes its natural timing period, the period will end regeneratively in the way already described, but if the trigger signal is still present at this moment the timing cycle will terminate non-regeneratively, and the output pulse will have a longer period and fall-time than in the former case.

Electronic triggering

Figures 4.14 and *4.15* show ways of applying electronic (rather than manual) triggering to the transistor monostable circuit. In each case, the circuit is triggered by a square-wave input signal with a short rise time; this waveform is differentiated by C_2–R_6, to produce a brief trigger pulse. In the *Figure 4.14* circuit the differentiated input signal is discriminated by diode D_1, to provide a positive trigger pulse to Q_2 base each time an external trigger signal is applied. In the *Figure 4.15*

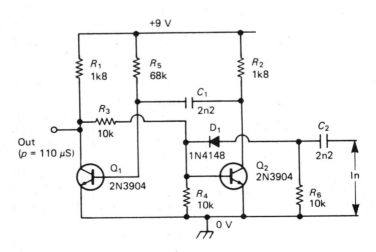

Figure 4.14 *Electronically-triggered monostable*

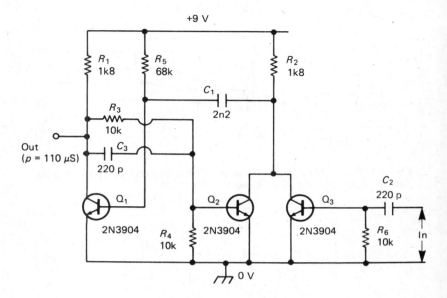

Figure 4.15 *Monostable with gate-input triggering*

circuit the differentiated trigger signal is fed to the monostable via *gate transistor* Q_3. Note in the latter design that speed-up capacitor C_3 is wired across feedback resistor R_3, to help improve the shape of the circuit's output pulse.

The circuits in *Figures 4.14* and *4.15* each give an output pulse period of about 110 μs with the component values shown. The period can be varied from a fraction of a microsecond to several seconds by suitable choice of the C_1–R_5 values.

4001B/4011B CMOS monostable circuits

One of the cheapest and easiest ways of making a standard or a resettable monostable is to use a CMOS 4001B quad 2-input NOR gate or a 4011B quad 2-input NAND gate IC in one of the configurations shown in *Figures 4.16* to *4.19*. Note, however, that the output pulse widths of these circuits are subject to fairly large variations between individual ICs and with variations in supply rail voltage, and these circuits are thus not suitable for use in high-precision applications.

Figures 4.16 and *4.17* show alternative versions of the standard

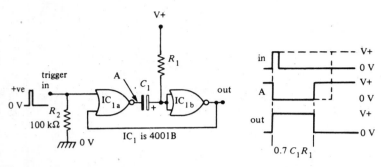

Figure 4.16 *CMOS 2-gate NOR monostable is triggered by a positive-going signal and generates a positive-going output pulse*

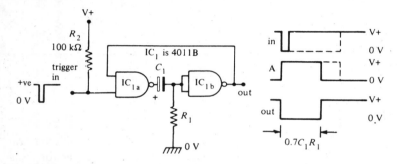

Figure 4.17 *CMOS 2-gate NAND monostable is triggered by a negative-going signal and generates a negative-going output pulse*

monostable circuit, each using only two of the four available gates in the specified CMOS package. In these circuits the duration of the output pulse is determined by the C_1–R_1 values, and approximates $0.7 \times C_1 \times R_1$. Thus, when R_1 has a value of 1M5 the pulse period approximates 1 s per µF of C_1 value. In practice, C_1 can have any value from about 100 pF to a few thousand microfarads and R_1 can vary from 4k7 to 10 M.

An outstanding feature of these circuits is that the input trigger pulse or signal can be direct coupled and its duration has little effect on the length of the generated output pulse. The NOR version of the circuit (*Figure 4.16*) has a normally-low output and is triggered by the edge of a positive-going input signal, and the NAND version (*Figure 4.17*) has a normally-high output and is triggered by the edge of a negative-going input signal.

Another feature is that the pulse signal appearing at point A has a period equal to that of either the output pulse or the input trigger pulse, whichever is the greater of the two. This feature is of value when making pulse-length comparators and over-speed alarms.

The operating principle of these monostable circuits is fairly simple. Look first at the case of the *Figure 4.16* circuit, in which IC_{1a} is wired as a NOR gate and IC_{1b} is wired as an inverter. When this circuit is in the quiescent state the trigger input terminal is held low by R_2, and the output of IC_{1b} is also low. Thus, both inputs of IC_{1a} are low, so IC_{1a} output is forced high and C_1 is discharged.

When the positive trigger signal is applied to the circuit the output of IC_{1a} is immediately forced low and (since C_1 is discharged at this moment) pulls IC_{1b} input low and thus drives IC_{1b} output high; IC_{1b} output is coupled back to the IC_{1a} input, however, and thus forces IC_{1a} output to remain low irrespective of the prevailing state of the trigger signal. As soon as IC_{1a} output switches low, C_1 starts to charge up via R_1 and, after a delay determined by the C_1–R_1 values, the C_1 voltage rises to such a level that the output of IC_{1b} starts to swing low, terminating the output pulse. If the trigger signal is still high at this moment, the pulse terminates non-regeneratively, but if the trigger signal is low (absent) at this moment the pulse terminates regeneratively.

The *Figure 4.17* circuit operates in a manner similar to that described above, except that IC_{1a} is wired as a NAND gate, with its trigger input terminal tied to the positive supply rail via R_2, and the R_1 timing resistor is taken to ground.

Resettable circuits

In the *Figures 4.16* and *4.17* circuits the output is direct-coupled back to one input of IC_{1a} to effectively maintain a trigger input once the true trigger signal is removed, thereby giving a semi-latching circuit operation. These circuits can be modified so that they act as **resettable monostables** by simply providing them with a means of breaking this feedback path, as shown in *Figures 4.18* and *4.19*.

Here, the feedback connection from IC_{1b} output to IC_{1a} input is made via R_3. Consequently, once the circuit has been triggered and the *original trigger signal has been removed* each circuit can be reset by forcing the feedback input of IC_{1a} to its normal quiescent state via push-button switch PB_1. In practice, PB_1 can easily be replaced by a

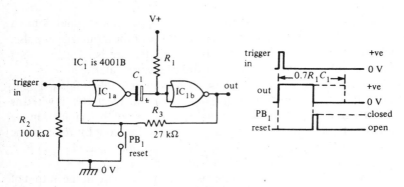

Figure 4.18 *Resettable NOR-type CMOS monostable*

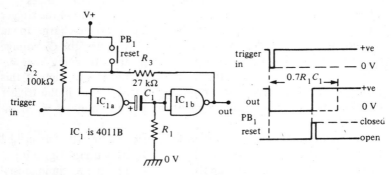

Figure 4.19 *Resettable NAND-type CMOS monostable*

transistor or CMOS switch, etc., enabling the reset function to be accomplished via a suitable reset pulse.

CMOS flip-flop monostables

Medium-accuracy monostables can easily be built by using standard edge-triggered CMOS flip-flop ICs such as the 4013B dual D-type or the 4027B dual JK-type in the configurations shown in *Figures 4.20* and *4.21*. Both of these circuits operate in the same basic way, with the IC wired in the *frequency divider* mode by suitable connection of its *control terminals* (data and set in the 4013B, and J, K, and set in the 4027B), but with the Q terminal connected back to reset via a C–R time-delay network. The operating sequence of each circuit is as follows.

When the circuit is in its quiescent state the Q output terminal is low and discharges timing capacitor C_1 via R_2 and the parallel combination D_1–R_1. On the arrival of a sharply rising leading edge on the clock terminal the Q output flips high, and C_1 starts to charge up via the series combination R_1–R_2 until eventually, after a delay determined mainly by the C_1–R_1 values (R_1 is large relative to R_2), the C_1 voltage rises to such a value that the flip-flop is forced to reset, driving the Q terminal low again. C_1 then discharges rapidly via R_2 and D_1–R_1, and

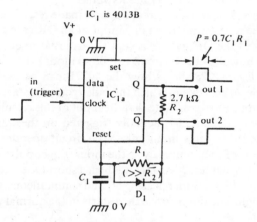

Figure 4.20 *D-type CMOS flip-flop used as a monostable*

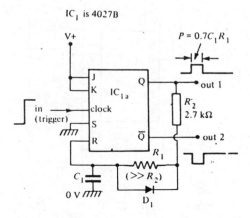

Figure 4.21 *JK-type CMOS flip-flop used as a monostable*

the circuit is then ready to generate another pulse on the arrival of the next trigger signal.

The timing period of these circuits are roughly equal to $0.7 \times C_1 \times R_1$, and the reset period (the time taken for C_1 to discharge at the end of each pulse) roughly equals $C_1 \times R_2$. R_2 is used mainly to prevent degradation of the trailing edge of the pulse waveform as C_1 discharges, and can be reduced to zero if this degradation is acceptable. Note that the circuit generates a positive-going pulse at the Q output, and a negative-going one at the not-Q output, and that the not-Q waveform is not influenced by the R_2 value.

The *Figures 4.20* and *4.21* circuits can be made resetable by connecting C_1 to the reset terminal via one input of an OR gate and using the other input of the OR gate to accept the external reset signal. *Figure 4.22* shows how the 4027B circuit can be so modified.

Finally, *Figure 4.23* shows how the 4027B can be used to make a **retriggerable monostable** in which the pulse period restarts each time a new trigger pulse arrives. Note that the input of this circuit is normally high, and that the circut is actually triggered on the trailing (rising) edge of a negative-going input pulse. The circuit operates as follows.

At the start of each timing cycle the input trigger pulse switches low and rapidly discharges C_1 via D_1 and then, a short time later, the trigger pulse switches high again, releasing C_1 and simultaneously flipping the Q output high. The timing cycle then starts in the normal way, with C_1

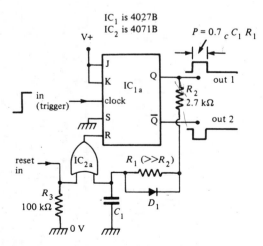

Figure 4.22 *Resettable JK-type CMOS monostable*

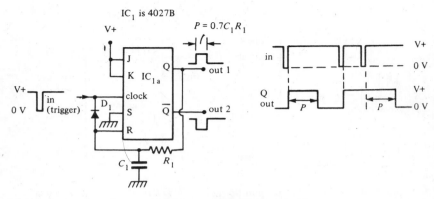

Figure 4.23 *Retriggerable JK-type CMOS monostable*

charging via R_1 until the C_1 voltage rises to such a level that the flip-flop resets, driving the Q output low again and slowly discharging C_1 via R_1.

If a new trigger pulse arrives in the midst of a timing period (when Q is high and charging C_1 via R_1), C_1 discharges rapidly via D_1 on the low part of the trigger, and commences a new timing cycle as the input waveform switches high again. In practice, the input trigger pulse must be wide enough to fully discharge C_1, but should be narrow relative to the width of the output pulse. The timing period of the output pulse equals $0.7 \times C_1 \times R_1$. For best results, R_1 should have as large a value as possible.

4047B and 4098B CMOS monostables

A number of dedicated CMOS and TTL monostable pulse generator ICs are available, and are well worth using in some circuit applications. The best known of the CMOS devices are the 4047B monostable/astable IC, and the 4098B dual monostable (a greatly improved version of the 4528B). *Figure 4.24* shows the outlines and pin notations of these two devices.

Note that, like the other CMOS-based monostables that we have already looked at, the 4047B and 4098B have rather poor pulse-width accuracy and stability. These ICs are, however, quite versatile, and can be triggered by either the positive or the negative edge of an input signal, and can be used in either the standard or the retriggerable mode.

The 4047B actually houses an astable multi and a frequency-divider stage, plus logic networks. When used in the monostable mode the

80 Timer/Generator Circuits Manual

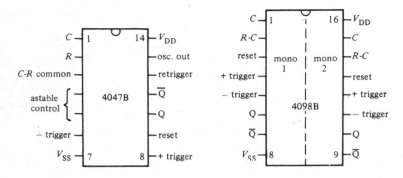

Figure 4.24 *Outlines and pin notations of the 4047B monostable/astable and 4098B dual monostable CMOS ICs*

trigger signal starts the astable and resets the counter, driving its Q output high. After a number of *C–R* controlled astable cycles the counter flips over and simultaneously kills the astable and switches the Q output low, completing the operating sequence. Consequently, the *C–R* timing components produce relatively long output pulse periods, this period approximating $2.5 \times C \times R$.

In practice, *R* can have any value from 10 k to 10 M. *C* must be a non-polarized capacitor with a value greater than 1 nF. *Figure 4.25(a)* and *(b)* shows how to connect the IC as a standard monostable triggered by either positive or negative input edges, and *Figure 4.25(c)* shows how to connect the monostable in the retriggerable mode. Note that these circuits can be reset at any time by pulling reset pin 9 high.

The 4098B is a fairly simple dual monostable, in which the two mono sections share common supply connections but can otherwise be used independently. Mono 1 is housed on the left side (pins 1 to 7) of the IC, and mono 2 on the right side (pins 9 to 15) of the IC. The timing period of each mono is controlled by a single resistor (*R*) and capacitor (*C*), and approximates $0.5 \times C \times R$. *R* can have any value from 5k0 to 10 M, and *C* can have any value from 20 pF to 100 µF. *Figure 4.26* shows a variety of ways of using the 4098B. Note in these diagrams that the bracketed numbers relate to the pin connections of mono 2, and the plain numbers to mono 1, and that the reset terminal (pins 3 or 13) is shown disabled.

Figure 4.26(a) and *(b)* shows how to use the IC to make retriggerable monostables that are triggered by positive or negative input edges respectively. In *Figure 4.26(a)* the trigger signal is fed to the +trig pin

Pulse generator circuits

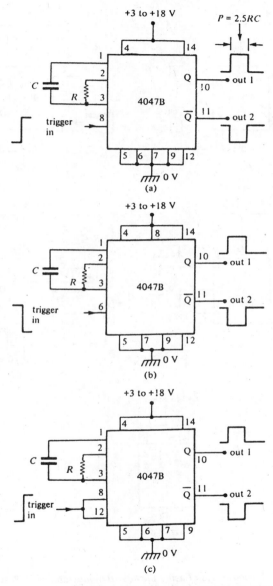

Figure 4.25 *Various ways of using the 4047B as a monostable. (a) Positive edge-triggered monostable. (b) Negative edge-triggered monostable. (c) Retriggerable monostable, positive edge-triggered*

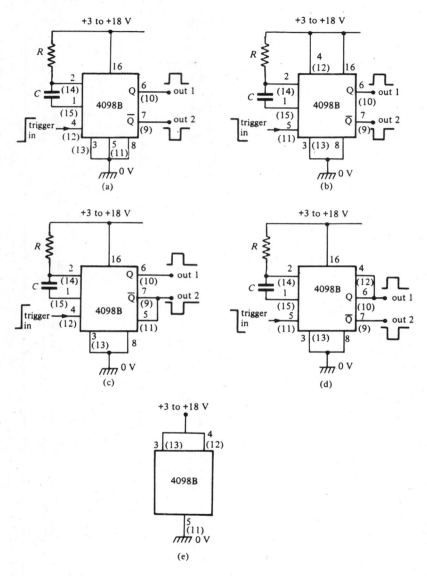

Figure 4.26 *Various ways of using the 4098B monostable. (a) Positive edge-triggering, retriggerable mono. (b) Negative edge-triggering, retriggerable mono. (c) Positive edge-triggering (non-retriggerable mono). (d) Negative edge-triggering (non-retriggerable mono). (e) Connections for each unused section of the IC*

and the −trig pin is tied low. In *Figure 4.26(b)* the trigger signal is applied to the −trig pin and the +trig pin is tied high.

Figure 4.26(c) and *(d)* shows how to use the IC to make standard (non-retriggerable) monostables that are triggered by positive or negative edges respectively. These circuits are similar to those mentioned above except that the unused trigger pin is coupled to either the Q or the not-Q output, so that trigger pulses are blocked once a timing cycle has been initiated.

Finally, *Figure 4.26(e)* shows how the unused half of the IC must be connected when only a single monostable is wanted from the package. The −trig pin is tied low, and the +trig and reset pins are tied high.

The 74121N TTL monostable

The 74121N is a dedicated TTL monostable pulse generator IC that can usefully generate output pulse widths from a few tens of nanoseconds up to several hundred milliseconds. *Figure 4.27* shows the outline, pin notations, and simplified internal circuit of the device, which can be triggered on either the leading or trailing edges of an input waveform, and has three alternative input trigger terminals.

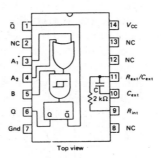

Figure 4.27 *Outline, pin notations, and simplified internal circuit of the 74121 TTL monostable multi-vibrator IC*

Dealing first with methods of triggering the IC, *Figures 4.28* and *4.29* show basic ways of using it when it is connected as a simple 30 ns pulse generator using its built-in timing components. In the *Figure 4.28* circuit the trigger signal is applied to pin 5 B input terminal via a transistor buffer stage. This pin is connected to an internal Schmitt gate that triggers the monostable on the leading edge of the input waveform if

84 Timer/Generator Circuits Manual

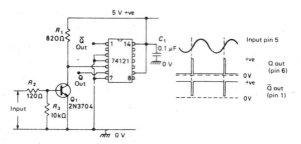

Figure 4.28 *30 ns TTL pulse generator using B input leading-edge triggering*

pins 3 (A_1) and/or 4 (A_2) are low (at logic-0) at this time. Note that slow-rising edges can be used to trigger the monostable via this pin 5 input terminal.

Figure 4.29 shows the IC wired so that it triggers on the falling or trailing edge of the input waveform. Here, the input signal is fed to pins 3 (A_1) and 4 (A_2) of the IC via a 7413 TTL Schmitt IC, which generates the sharp edges needed to operate these inputs. A_1 and A_2 are negative-edge triggered logic inputs, and trigger the monostable when either goes to logic-0 with input B disconnected or biased at logic-1.

Dealing next with the IC's timing circuitry, *Figures 4.30* to *4.32* show different ways of achieving this. The IC has three timing-component terminals. A low-value timing capacitor is built into the IC, and can be augmented by external capacitors wired between pins 10 and 11 (positive terminals of polarized capacitors should go to pin 11). The IC also houses a 2k0 resistor, which can be used as a timing component by wiring pin 9 to pin 14, either directly or via an external series resistor (maximum value 40 k); alternatively, the internal resistor can be

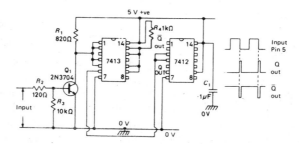

Figure 4.29 *30 ns TTL pulse generator using A_1 and A_2 input trailing-edge triggering*

ignored and an external timing resistor (1k4 to 40k) can be wired between pins 11 and 14. Whichever connection is used, the output pulse width = 0.7 $R_T C_T$, where width is in milliseconds, R_T is the total timing resistance in kilohms, and C_T is the timing capacitance in microfarads.

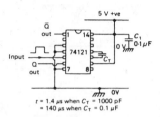

Figure 4.30 *TTL pulse generator using internal timing resistor and external timing capacitor*

Thus, in *Figure 4.30*, timing is achieved using the IC's internal 2k0 resistor plus external capacitor C_T. In *Figure 4.31*, timing is achieved using both internal and external resistors, plus an external capacitor, and in *Figure 4.32* it is achieved using external components only. Note, incidentally, that the circuits of *Figure 4.28* and *4.29* each use the IC's internal timing components only, and give pulse periods of about 30 ns.

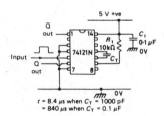

Figure 4.31 *TTL pulse generator using internal and external timing resistors and external timing capacitor*

The basic *Figure 4.32* circuit is of great value in decade-ranged variable pulse generator applications, and *Figures 4.33* shows how it can be adapted to act as a high-performance add-on pulse generator that covers the range 100 ms to 100 ns in six decade ranges.

Finally, to complete this chapter, *Figure 4.34* shows how two of the above circuits can be coupled together to make an add-on wide-range delayed pulse generator, which does not generate its final output pulse

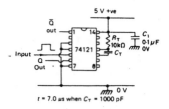

Figure 4.32 *TTL pulse generator using external timing resistor and capacitor*

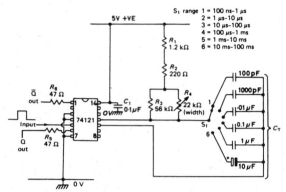

Figure 4.33 *High-performance add-on TTL pulse generator covers 100 ns to 100 ms*

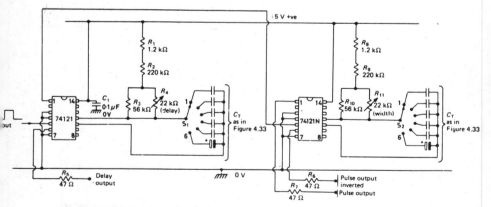

Figure 4.34 *High-performance add-on TTL delayed pulse generator covers 100 ns to 100 ms*

until some pre-set time after the arrival of the initial trigger pulse. Note that this circuit (and that of *Figure 4.32*) gives both inverted and non-inverted outputs, which are each of fixed amplitude and are short-circuit protected via 47 Ω series resistors.

5 Timer IC generator circuits

Timer ICs are designed to generate accurate and stable $C-R$ defined timing periods, for use in a variety of monostable pulse generator and astable square-wave generator applications. The best known of these is the highly versatile 555 family of devices, which are available in both *single* (555) and *dual* (556) bipolar packages and also in CMOS forms (7555 and 7556), and these are the main types of timer IC described in this chapter. Two other types of timer ICs are also briefly described, however, these being the ZN1034 long-period timer and the 2240 programmable timer, which are both of value in applications where very long timing periods are needed.

555 basics

The 555 timer IC was first introduced by Signetics in the late 1970s but is now produced by many other IC manufacturers. The 555 has many attractive features: it can operate from a wide range of supply voltages; can source (supply) or sink (absorb) fairly high load currents; and consumes fairly modest supply currents. It can be used as either a monostable or astable multivibrator, with its timing periods variable from a few microseconds to hundreds of seconds via a simple $C-R$ network; timing stability is almost independent of variations in supply rail voltage and in temperature.

The 555 produces excellent output waveforms, with typical rise and fall times of about 100 ns. When used in the monostable mode its output can be pulse-width modulated (PWM) if required, and when used in the astable mode its waveform can easily be subjected to frequency-sweep control, to frequency modulation (FM), or to pulse-position modulation (PPM).

The 555 is available under a wide variety of manufacturer's type numbers, but is generally known simply as a **555 timer** if it uses bipolar construction or as a **7555** or **CMOS 555** if it uses a CMOS form of construction. It is available in both 8 pin DIL and 8 pin TO-99 packages, as shown in *Figures 5.1(a)* and *(b)*. The dual (556 or 7556) version of the IC is housed in the 14 pin DIL package shown in *Figure 5.2*. The table of *Figure 5.3* lists typical basic parameter values of the 555 and 7555 *single* versions of the device.

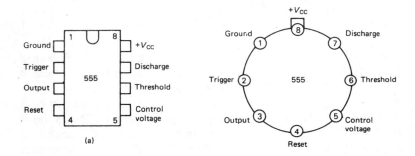

Figure 5.1 *Outline and pin notations of the (a) 8-pin DIL and (b) 8-pin TO-99 versions of the 555 (or 7555) timer IC*

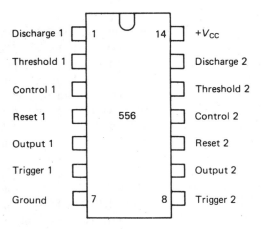

Figure 5.2 *Outline and pin notations of the 14-pin DIL version of the 556 (or 7556) dual timer IC*

Timer IC generator circuits 89

Parameter	555	7555
Supply voltage range	4.5–16 V	2–18 V
Power dissipation (max.)	600 mW	200 mW
Supply current (at V_{CC} = 15 V)	10 mA	100 µA
Max output source/sink current	200 mA	100 mA
Timing accuracy, typical	± 1%	± 2%
Drift with temperature	50 ppm/°C	50 ppm/°C
Drift with supply voltage	0.1%/V	1%/V
Threshold voltage	2/3 V_{CC}	2/3 V_{CC}
Trigger voltage	1/3 V_{CC}	1/3 V_{CC}
Reset voltage	0.7 V	0.7 V
Output rise/fall times	100 ns	40 ns

Figure 5.3 *555 and 7555 parameter values*

How it works

Figure 5.4 shows (within the double lines) the functional diagram of the basic 555 timer IC, which is internally arranged in the form of two voltage comparators, one *R–S* flip-flop, a low-power complementary output stage, a *slave* transistor, and a supply-driven 3 × 5k0 potential divider that generates a 1/3 V_{cc} reference voltage that is fed to the non-inverting input of the lower comparator and a 2/3 V_{cc} reference that is fed to the inverting input of the upper comparator. The outputs of these comparators control the *R–S* flip-flop, which in turn controls the states of the output stage and the slave transistor. The flip-flop state can also be controlled via the IC's pin 4 reset. Note that *Figure 5.4* also shows the connections for using the 555 as a basic monostable multi-vibrator or *timer*, and the following explanation assumes that the IC is connected in this configuration.

When the *Figure 5.4* timer circuit is in its quiescent state the pin 2 trigger terminal is held high via R_4, Q_1 is saturated and forms a short across timing capacitor C_T, and the pin 3 output terminal is driven low. The monostable timer action is initiated by feeding a negative-going trigger pulse to pin 2, and as this pulse falls below the internal 1/3 V_{cc} reference value the output of the lower voltage comparator changes state and makes the *R–S* flip-flop switch over, turning Q_1 off and driving the pin 3 output high.

As Q_1 turns off it removes the short from C_T, so C_T starts to charge exponentially via R_T until eventually the C_T voltage rises to 2/3 V_{cc}. At this point the IC's upper voltage comparator changes state and switches the *R–S* flip-flop back to its original state, turning Q_1 on and

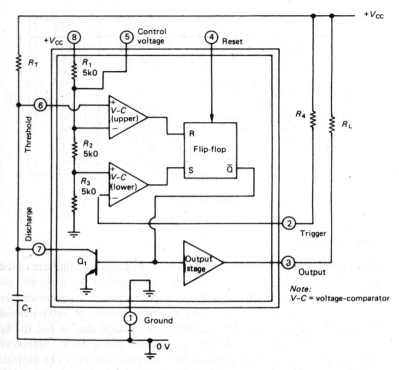

Figure 5.4 *Functional block diagram (within the dotted lines) of the 555 timer IC, with external connections for use as a timer*

rapidly discharging C_T and simultaneously switching output pin 3 low again, thus completing the operating sequence.

Note that, once triggered, this circuit cannot respond to additional triggering until the timing sequence is complete, but that the sequence can be aborted at any time by feeding a negative-going pulse to reset pin 4. The timing period of the circuit, in which the pin 3 output is high, is given as:

$$t = 1.1\, R_T\, C_T$$

where t is in milliseconds, R_T is in kilohms and C_T is in microfarads.

The graph of *Figure 5.5* shows how delays of 10 μs to 100s can be obtained by selecting suitable C_T and R_T values in the range 1 nF to 100μF and 1k0 (minimum permitted value) to 10 M. Note that C_T

must be a low leakage component, and that the timing period is virtually independent of supply voltage value but can be varied by applying a variable resistance or voltage between the ground and the pin 5 control voltage terminal of the IC; this facility enables the periods to be externally modulated or compensated.

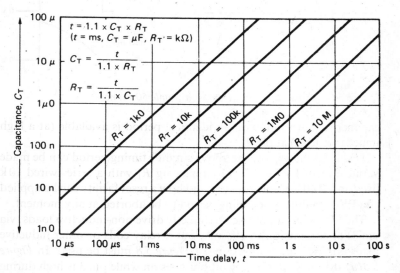

Figure 5.5 555 time delays (t) for various values of R_T and C_T

The IC's pin 3 output terminal is normally low, but switches high during the monostable timing period, and can source or sink currents (up to 200 mA in the bipolar 555), so external loads can be connected between pin 3 and either the positive supply rail or ground, depending on the desired type of load operation.

Practical timer circuits

Figure 5.6 shows the practical circuit of a simple fixed-period (about 50 s) manually-triggered 555 timer, together with relevant circuit waveforms. The circuit is similar to that of *Figure 5.4*, except that the timing action is initiated by briefly closing start switch PB_1, that pin 5 is decoupled via C_2, and that the output state is visible via a LED. The fixed-period output pulse (set via R_1–C_1) is available at pin 3, and an

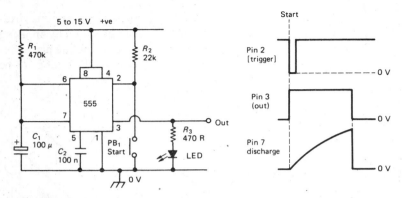

Figure 5.6 *Circuit and waveforms of a simple 50 s timer*

exponential sawtooth with an identical period is available (at a high impedance level) at pin 7.

Figure 5.7 shows how the above circuit's timing period can be made variable from 1.1 s to 120 s by replacing R_1 with a series-wired 10 k fixed and 1M0 variable resistor, and how a reset facility can be applied (via PB_1), enabling the timing period to be aborted at any moment.

The 555 timer can be used to directly drive non-inductive loads (via pin 3) at currents up to 200 mA, but if relay coils or other inductive loads are used the connections of *Figure 5.8* must be used. In *Figure 5.8(a)* the relay is normally off but goes on while pin 3 is high during

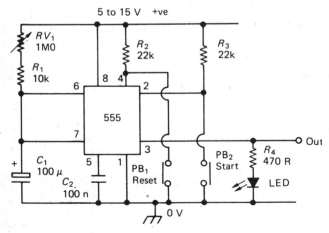

Figure 5.7 *1.1 s to 120 s timer with reset facility*

Timer IC generator circuits 93

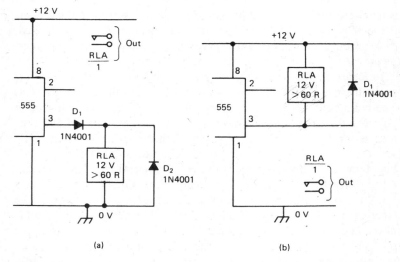

Figure 5.8 *Alternative ways of driving a relay from the output of a 555*

the timing interval; in *Figure 5.8(b)* the relay is normally on but turns off during the timing interval. In these circuits the diodes protect the 555 against inductive-switching damage; relay contacts RLA/1 can be used to control external circuitry.

Figure 5.9 shows a simple relay-output timer that spans the range 1.1 s to 120 s in two switch-selected decade ranges. This is a useful

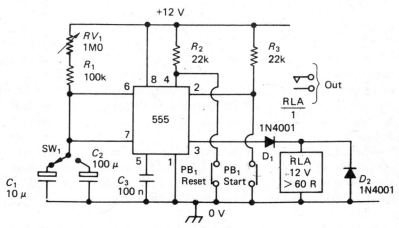

Figure 5.9 *Simple two-range, 1.1 s to 120 s, relay-output timer*

general-purpose circuit, but it consumes current even when the timer is in the off mode, and its two RV_1 scales must be individually calibrated, since timing capacitors C_1 and C_2 are wide tolerance electrolytic types. *Figure 5.10* shows how these two defects can be overcome.

In *Figure 5.10*, power is fed to the timer circuit via PB_1 or RLA/1; normally, these are both open, so the circuit consumes zero current. The timing cycle is initiated by briefly closing PB_1, thus connecting power to the circuit. At this moment, C_3 is fully discharged and thus feeds a start pulse to the 555's pin 2 via R_4, thus starting the timing cycle and driving relay RLA on and making contacts RLA/1 close, thus maintaining the circuit's power connection even when PB_1 is released. At the end of the timing cycle the relay turns off again and contacts RLA/1 re-open, thus removing power from the timer circuit again. The circuit's timing is controlled mainly by R_1–RV_1 and by C_1 or C_2, but is also influenced by the settings of RV_2 and RV_3, which connect to pin 5 of the IC and enable the timing to be trimmed so that the two timing ranges can use a single calibrated scale, even though wide-tolerance timing capacitors are used.

To set up the *Figure 5.10* circuit, first set RV_1 to maximum value, set range switch SW_1 to position 1, activate start button PB_1, and adjust RV_2 to give a timing period of precisely 10 s. Next, set SW_1 to position 2, activate PB_1, and adjust RV_3 to give a timing period of 100 s. Adjustments are then complete, and the timing scale can be calibrated over the full 10 s range.

In-car timer circuits

Figure 5.11 shows a 555 timer used as an automatic delayed-turn-off headlight control system for use in automobiles. This unit lets the owner use the car lights to illuminate his path for a pre-set period after parking and leaving the car; the system does not interfere with normal headlight operation under actual driving conditions. It works as follows.

When the vehicle's ignition switch is on, current is fed to the relay coil via D_3, so RLA is on and contacts RLA/1 are closed, connecting the 12 V supply to both timer circuit and the headlight switch. The headlights thus operate in the normal way under this condition, but C_2 is fully discharged.

When the ignition is first switched off the relay tries to open, but at the same instant a negative trigger pulse is fed to the timer via C_2 and

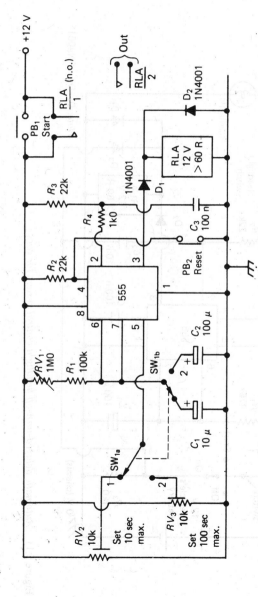

Figure 5.10 *Precision (compensated) two-range (0.9 s to 10 s; 9 s to 100 s) timer*

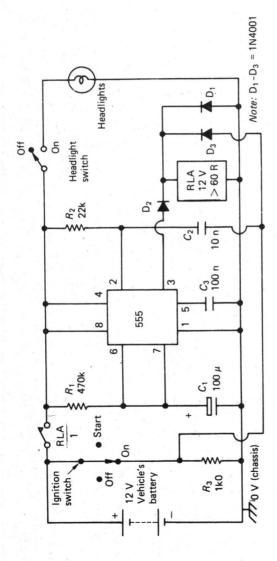

Figure 5.11 *Automatic delayed-turn-off headlight control system for cars*

initiates a 50 s timing cycle that feeds current to the relay coil via D_2, thus maintaining RLA/1's connection to the headlight switch for 50 s after the ignition is turned off. At the end of this period the relay turns off and contacts RLA/1 open, breaking the supply connection to the timer circuit and the headlight switch, and the operation is complete.

Note that the above mode of circuit operation is compatible with the lighting system used in most modern vehicles, in which the headlight switch is fed via the ignition switch. On older types of vehicle, where headlight operation is independent of the ignition switch, a manually-triggered delayed turn-off headlight or spotlight control facility can be obtained by using the circuit of *Figure 5.12*. The action here is such that, if the vehicle has its lights off, they can be turned on for a pre-set 50 s period by briefly pressing a start switch.

This circuit uses a relay with two sets of NO contacts, and the timing sequence is initiated by briefly closing PB_1. Normally, both PB_1 and the relay contacts are open, so no power reaches the timer, and the lights are off; C_2 is discharged. When PB_1 is briefly closed the relay turns on, and contacts RLA/1 connect power to the timer circuit and simultaneously C_2 feeds a trigger pulse to the 555 timer and initiates a 50 s timing cycle, and simultaneously contacts RLA/2 close and connect power to the vehicle's light. At the end of the timing cycle the relay turns off again and removes power from both the lights and the timer circuit, and the operation is complete.

An automatic porch light

To complete this look at timer applications of the 555, *Figure 5.13* shows how the IC can be used to make a smart timer that automatically turns a porch light on for 50 s when the presence of a visitor is detected, but does so only under dark conditions. The presence of the visitor is detected via SW_1, which can take the form of a microswitch activated by a porch gate, or a pressure-pad switch that is activated by body weight, and the dark condition is detected by a light-sensitive resistor (LDR).

Circuit operation relies on the fact that timer triggering can only occur if the IC's pin 2 trigger pulse falls below the $1/3$ V_{cc} value, and in *Figure 5.13* the pulse is generated by closing SW_1 but the pulse magnitude is controlled by the LDR–RV_1 potential divider and is dependent on light level. Thus, under bright conditions the LDR has a low value and the LDR–RV_1 junction voltage is high, so effective trigger pulses

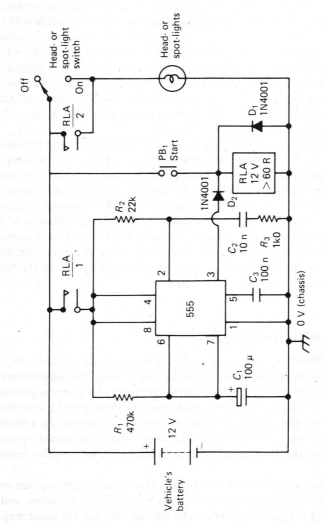

Figure 5.12 *Manually-triggered, delayed turn-off light control system for cars*

Timer IC generator circuits 99

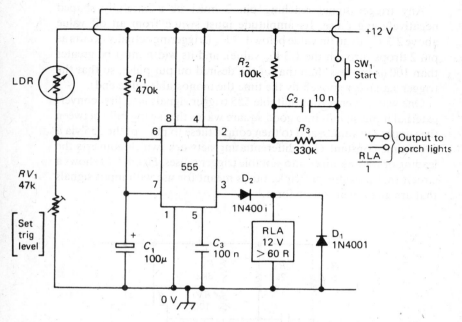

Figure 5.13 *Automatic porch light turns on for a preset period only when triggered at night*

cannot be generated, but under dark conditions the LDR resistance is high and the LDR–RV_1 junction voltage is low, and effective trigger pulses are generated each time that SW_1 is closed. The LDR can be any cadmium-sulphide photocell that presents a resistance in the range 1k0 to 47 k at the required minimum dark turn-on state, and RV_1 can be used to preset the minimum dark level at which the circuit will trigger.

Pulse generator circuits

The 555 circuits shown so far all act as manually-triggered monostable multivibrators or pulse generators. The 555 can be used as a conventional electronically-triggered monostable pulse generator by feeding suitable trigger signals to pin 2 and taking the output pulse signals from pin 3. The IC can be used to generate good output pulses with periods from 5 μs to hundreds of seconds. The maximum useful pulse repetition frequency is about 100 kHz.

Any trigger signal reaching pin 2 must be a carefully shaped negative-going pulse. Its amplitude must switch from an off value above $2/3\ V_{cc}$ to an on value below $1/3\ V_{cc}$ (triggering actually occurs as pin 2 drops through the $1/3\ V_{cc}$ value), and its width must be greater than 100 ns but less than that of the desired output pulse, so that the trigger signal is removed by the time the monostable pulse ends.

One way of generating suitable 555 trigger signals is to first convert external input signals into good square waves that swing fully between the supply rail values, and to then couple these to pin 2 of the 555 via a short time-constant C–R differentiating network, which converts the leading or trailing edges into suitable trigger pulses. *Figure 5.14* shows a circuit that uses this principle, but is meant for use with input signals that are already in square form.

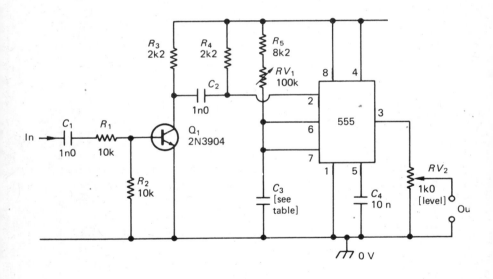

C_3 value	Pulse width range
10 µF	90 ms–1.2 s
1 µF	9 ms–120 ms
100 nF	900 µs–12 ms
10 nF	90 µs–1.2 ms
1 nF	9 µs–120 µs

Figure 5.14 *Simple add-on pulse generator is triggered by rectangular input signals*

Here, Q_1 converts the input signal into one that switches fully between the supply rail values, and these are fed to pin 2 via the C_2–R_4 differentiating network. This circuit can be used as an add-on pulse generator in conjunction with an existing square-wave generator. Variable-amplitude output pulses are available via RV_2, and their widths are variable over a decade range via RV_1 and can be switched in decade ranges by using the C_3 values shown in the table; the total pulse width range spans 9 μs to 1.2 s. C_4 decouples pin 5 and improves circuit stability.

Figure 5.15 shows how the above circuit can be modified so that it can be directly driven by any type of input, including a sine wave. Here, IC_1 is wired as a Schmitt trigger and converts all input signals into a rectangular form that is used to drive the IC_2 monostable in the same way as described above. This circuit can be used as an add-on pulse generator in conjunction with any free-running generator that gives peak-to-peak outputs greater than 1/2 V_{cc}.

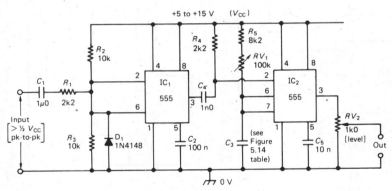

Figure 5.15 *Improved add-on pulse generator is triggererd by any input waveform*

Finally, to complete this look at 555 pulse generator circuits, *Figure 5.16* shows how three 555 ICs can be used to make an add-on delayed-pulse generator, in which IC_1 is used as a Schmitt trigger, IC_2 is a monostable that is used to control the pulse's delay width, and IC_3 is used as the final pulse generator. The final output pulse appears some delayed time (set via IC_2) after the application of the initial input trigger signal.

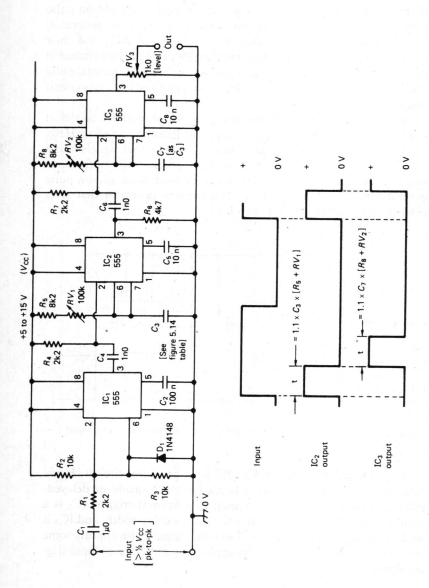

Figure 5.16 *Add-on delayed-pulse generator is triggered by any input waveform*

555 astable operation

The 555 timer IC can be used as a free-running astable multivibrator or square-wave generator by using it in the basic configuration shown in *Figure 5.17*, in which trigger pin 2 is shorted to the pin 6 threshold terminal, and timing resistor R_2 is wired between pin 6 and discharge pin 7. To understand the circuit operation, relate the following explanation to the 555 functional block diagram of *Figure 5.4*.

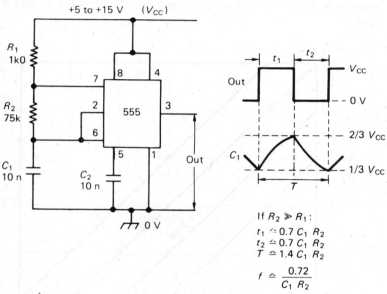

If $R_2 \gg R_1$:
$t_1 \simeq 0.7 \, C_1 \, R_2$
$t_2 \simeq 0.7 \, C_1 \, R_2$
$T \simeq 1.4 \, C_1 \, R_2$

$$f \simeq \frac{0.72}{C_1 \, R_2}$$

Figure 5.17 *Basic 1 kHz 555 astable multivibrator*

When power is first applied to this circuit C_1 starts to charge exponentially (in the normal monostable fashion) via the series R_1–R_2 combination, until eventually the C_1 voltage rises to 2/3 V_{cc}. At this point the basic monostable action terminates and discharge pin 7 switches to the low state. C_1 then starts to discharge exponentially into pin 7 via R_2, until eventually the C_1 voltage falls to 1/3 V_{cc}, and trigger pin 2 is activated. At this point a new monostable timing sequence is initiated, and C_1 starts to recharge towards 2/3 V_{cc} via R_1 and R_2. The whole sequence then repeats *ad infinitum*, with C_1 alternately charging towards 2/3 V_{cc} via R_1–R_2 and discharging towards 1/3 V_{cc} via R_2 only.

Note in the above circuit that when R_2 is very large relative to R_1 the operating frequency is determined mainly by R_2 and C_1 and that an almost symmetrical square-wave output is developed on pin 3 and a near-linear triangle waveform appears across C_1 under this condition. The graph of *Figure 5.18* shows the consequent relationship between frequency and the C_1–R_2 values. In practice, the circuit's R_1 and R_2 values can be varied from 1k0 to tens of megohms; note, however, that R_1 affects the circuit's current consumption, since pin 7 is effectively grounded during half of each cycle. Also note that the waveform's duty cycle or M/S ratio can be varied by suitable choice of the R_1 and R_2 ratios.

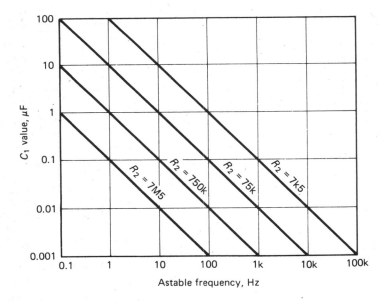

Figure 5.18 *Relationship between C_1, R_2, and 555 astable frequency when R_2 is large relative to R_1*

Figure 5.19 shows how the operating frequency of the *Figure 5.17* circuit can be made variable by simply replacing R_2 with a series-wired fixed and a variable resistor. With the component values shown the frequency can be varied from about 650 Hz to 7.2kHz via RV_1; the frequency span can be further increased by selecting alternative values of C_1.

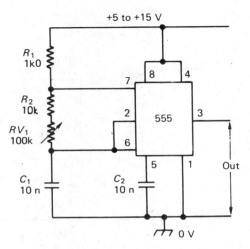

Figure 5.19   *Variable-frequency (650 Hz–7.2 kHz) square-wave generator*

Mark/space control

In each operating cycle of the *Figure 5.17* circuit C_1 alternately charges via R_1–R_2 and discharges via only R_2. Consequently, the circuit can be made to generate a non-symmetrical waveform with any desired mark/space (M/S) ratio by suitably selecting the R_1 and R_2 values. *Figures 5.20* to *5.23* show ways of making the M/S ratios fully variable.

Figures 5.20 and *5.21* show ways of achieving independent control of the mark and space periods. In *Figure 5.20*, C_1 alternately charges via R_1–D_1 and RV_1 and discharges via RV_2–D_2 and R_2. In *Figure 5.21*, C_1 alternately charges via R_1–RV_1 and D_1 and discharges via RV_2–D_2 and R_2. In both cases R_2 protects the IC against damage when RV_2 is reduced to zero, and the mark and space periods can each be independently varied over a 100:1 range, enabling the M/S ratio to be varied from 100:1 to 1:100; the frequency varies as the M/S ratio is altered.

Figures 5.22 and *5.23* show ways of altering the M/S ratio without significantly altering the operating frequency. In these circuits the mark period automatically increases as the space period decreases, and vice versa, so the total period of each cycle is constant. The most important waveform feature of these circuits is the **duty cycle** or relationship between the on time and total period of each cycle, and in both cases this is variable from 1% to 99% via RV_1.

In *Figure 5.22*, C_1 alternately charges via R_1–D_1 and the upper half of

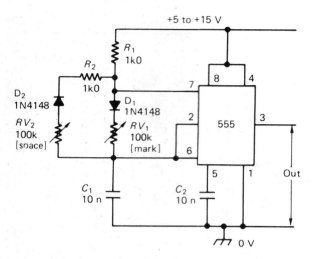

Figure 5.20 *Astable with mark and space periods independently variable from 7 µs to 750 µs*

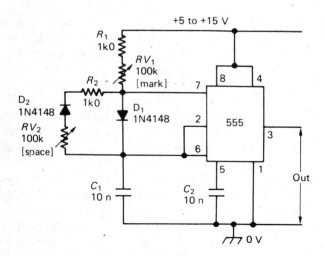

Figure 5.21 *Alternative version of the Figure 5.20 circuit*

Timer IC generator circuits 107

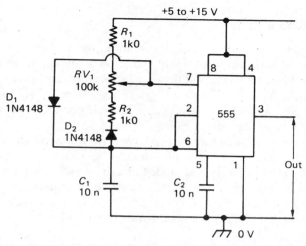

Figure 5.22 *1.2 kHz astable multi with duty cycle variable from 1% to 99%*

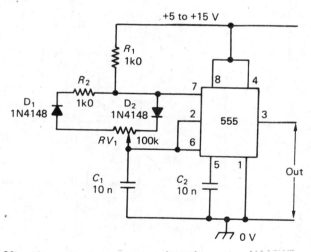

Figure 5.23 *Alternative version of the Figure 5.22 circuit*

RV_1 and discharges via D_2–R_2 and the lower half of RV_1. In *Figure 5.23*, C_1 alternately charges via R_1–D_1 and the right-hand half of RV_1 and discharges via D_2–R_2 and the left-hand half of RV_1. Both circuits operate at a nominal frequency of about 1.2 kHz with the C_1 value shown.

Precision astable

In the description of basic astable operation given earlier it was pointed out that in the initial half-cycle of operation timing capacitor C_1 charges from zero to $2/3\ V_{cc}$, but that in all subsequent half-cycles it either discharges from $2/3\ V_{cc}$ to $1/3\ V_{cc}$ or charges from $1/3\ V_{cc}$ to $2/3\ V_{cc}$. Consequently, the initial half-cycle of astable operation has a longer period than all subsequent half-cycles.

In some special applications this large period discrepancy may cause problems. In such cases the problem can be overcome by adding an external voltage divider and diode to bias C_1 to slightly below $1/3\ V_{cc}$ (rather than to zero volts) at the moment of switch-on, as shown in *Figure 5.24*. Here, R_1 rapidly charges C_1 to $1/3\ V_{cc}$ via D_1 at initial switch-on, but all C_1 charge is subsequently controlled by R_3 and/or R_4 only.

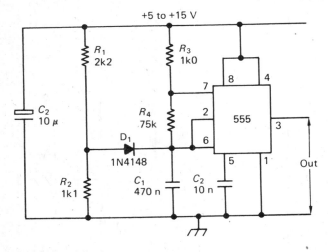

Figure 5.24 *Precision low-frequency (20 Hz approximately) astable*

Astable gating

The 555 astable can be gated on and off in several ways via either a switch or an electronic signal. One way is via the pin 4 reset terminal, and *Figures 5.25* and *5.26* show ways of gating the astable via this terminal and a push-button switch.

The action of the pin 4 reset terminal is such that the astable is enabled if pin 4 is biased above 0.7 V, but is disabled, with its output low, if pin 4 is pulled below 0.7 V by a current greater than 0.1 mA (by taking pin 4 to ground via 7k0 or less, for example). Thus, the *Figure 5.25* circuit is normally gated off by R_3, but can be turned on by closing PB$_1$ (thus biasing pin 4 high), and the *Figure 5.26* astable is normally on but can be gated off by closing PB$_1$ (thus shorting pin 4 to ground). These two circuits can also by gated by applying suitable electronic signals directly to pin 4.

Note in the *Figure 5.25* diagram that precise circuit waveforms are shown, and that the duration of the initial half-cycle is far longer than all others, and that C_1 takes a fairly long time to decay to zero when the astable is first gated off again. The *Figure 5.26* circuit in fact has similar characteristics.

Figure 5.27 shows another method of gating the 555 astable. Here, Q_1 is normally biased on via R_1 and thus acts like a closed switch that (via R_2) pulls the C_1–R_4 junction low and stops the astable operating, but when PB$_1$ is closed Q_1 is turned off and the astable operates in the normal way. Note that when the astable is gated on the initial half-cycle is again far longer than all others, and that the pin 3 output terminal is high when the astable is off.

Figure 5.28 shows the above circuit modified to give press-to-turn-off operation by replacing Q_1 with a push-button switch. Note that a digital signal can be used to gate this circuit by wiring a diode as shown and removing PB$_1$, in which case the circuit will turn off when the gate signal is below $1/3$ V_{cc}.

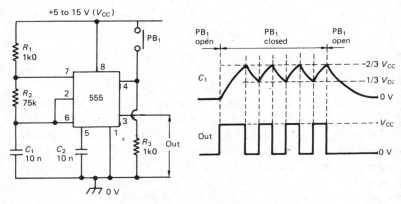

Figure 5.25 *Gated 1 kHz astable with press to turn on operation*

110 Timer/Generator Circuits Manual

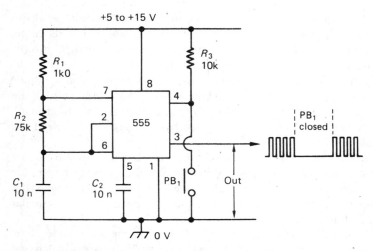

Figure 5.26 *Gated 1 kHz astable with press to turn off operation*

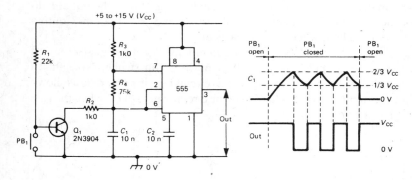

Figure 5.27 *Alternative gated 1 kHz astable with press to turn on operation*

To complete this look at gating techniques, *Figure 5.29* shows the *Figure 5.27* circuit modified to give precision operation. Here, when PB_1 is open Q_1 is saturated, so divider R_2–R_3 pulls the R_5–C_1 junction to just below $1/3$ V_{cc} via D_1, thus gating the astable off, but when PB_1 is closed Q_1 turns off and D_1 is reverse biased via R_2, and the astable is thus free to operate. Note that when PB_1 is first closed C_1 charges from an initial value of almost $1/3$ V_{cc} and the duration of the initial half-cycle is thus similar to all others.

Timer IC generator circuits 111

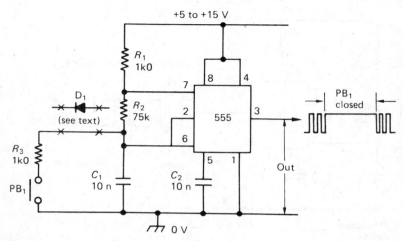

Figure 5.28 *Alternative gated 1 kHz astable with press to turn off operation*

Frequency modulation and pulse-position modulation

All the 555 astable circuits shown so far can be subjected to frequency modulation (FM) or to pulse-position modulation (PPM) by simply feeding the modulation signal to pin 5 (which connects to the IC's internal divider chain). This modulation signal may be an a.c. signal that is coupled to pin 5 via a blocking capacitor, as shown in *Figure 5.30*, or it may be a d.c. signal, as in *Figure 5.31*.

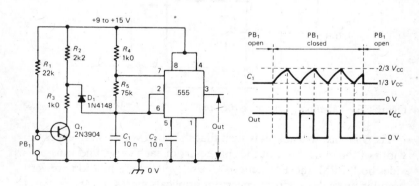

Figure 5.29 *Precision version of the Figure 5.27 circuit*

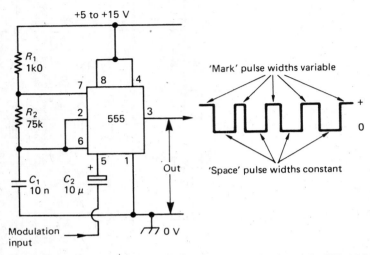

Figure 5.30 Method of applying a.c.-coupled FM or PPM to the 555 astable

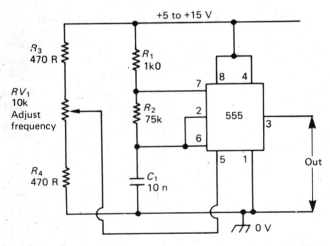

Figure 5.31 Method of applying d.c-coupled FM or PPM to the 555 astable

The 555's astable action is such that the pin 5 voltage influences the width of the mark but not the space part of each cycle, and thus provides both PPM and FM actions. These types of modulation are very useful in special waveform generator applications, as in various types of electronic siren and alarm-call generator circuit.

555 sirens and alarms

One very popular application of the 555 astable circuit is as a speaker-driving siren or alarm-call generator, and *Figures 5.32* to *5.37* show a selection of circuits of this type.

Figure 5.32 shows an 800 Hz monotone alarm-call generator circuit that can be used with a 5 to 15 V supply and with any speaker impedance. Note that Rx is wired in series with the speaker, to give a total load impedance of 75 Ω (to limit peak output currents to 200 mA). Available alarm output power depends on the speaker impedance and supply voltage values, i.e., it is 750 mW at 75 Ω at 15 V, etc.

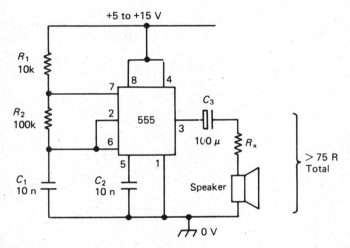

Figure 5.32 *Low-power (750 mW maximum) 800 Hz monotone alarm-call generator*

Figure 5.33 shows how to boost the output power of the above circuit to several watts via Q_1. Note that the resulting high output currents may modulate the supply voltage, and D_1 and C_3 help isolate the 555 from this modulation; D_2 and D_3 clamp the inductive switching spikes of the speaker and thus protect Q_1 from damage. This booster circuit is also used in the alarm circuits of *Figures 5.34* to *5.37*.

Figure 5.34 shows a pair of 555 astables used to make a pulsed-tone 800 Hz alarm-call generator. IC_1 acts as the 800 Hz generator, and is gated on and off once per second via IC_2 and D_1.

Figure 5.35 shows a warble-tone alarm-call generator that simulates the sound of a British police car siren. IC_1 is again wired as an alarm

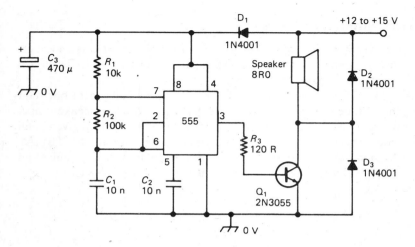

Figure 5.33 *Medium-power 800 Hz alarm-call generator*

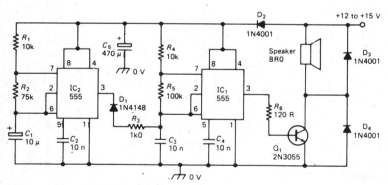

Figure 5.34 *Pulsed-tone (800 Hz) alarm-call generator*

tone generator and IC_2 as a 1 Hz astable, but in this case IC_2 output is used to frequency modulate IC_1 via R_5, the action being such that IC_1's frequency alternates between 440 Hz and 550 Hz at a 1 Hz cyclic rate.

Figure 5.36 shows a wailing alarm that simulates an American police car siren. Here, IC_2 is a low-frequency (6 s) astable that generates a ramp waveform that is buffered via Q_1 and used to frequency modulate tone generator IC_1 via R_6. IC_1 has a mid-frequency of about 800 Hz, and the modulation action is such that its output tone starts at a low frequency, rises for 3 s to a peak value, then falls back again for 3 s, and so on *ad infinitum*.

Timer IC generator circuits 115

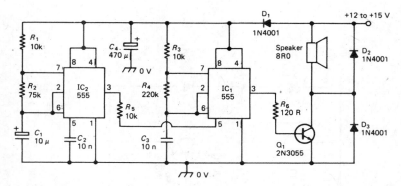

Figure 5.35 *Warble-tone alarm-call generator simulates British police car siren*

Finally, *Figure 5.37* shows an alarm that simulates the red alert sound used in *Star Trek* programmes. This sound starts at a low frequency, rises for about 1.15 s to a high tone, ceases for about 0.35 s, and then repeats *ad infinitum*. In this circuit IC_2 is a 1.5 s non-symmetrical astable that generates a rapidly rising but slowly falling sawtooth across C_1; this waveform is buffered via Q_1 and used (via R_7) to frequency modulate IC_1, making its frequency rise slowly during the falling parts of the sawtooth but collapse rapidly during the rising part. The rectangular pin 3 output of IC_2 gates IC_1 off via Q_2 during the collapsing part of the signal, so only the rising parts of the alarm signal are in fact heard.

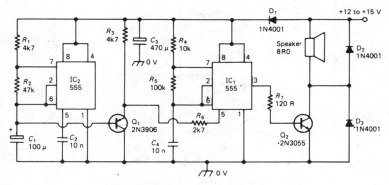

Figure 5.36 *Wailing alarm simulates American police siren*

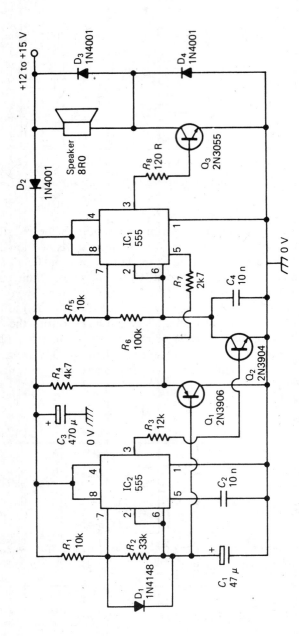

Figure 5.37 *Red alert siren simulates Star Trek alarm signal*

Long-period timers

An obvious weakness of the 555 timer circuit is that its accuracy decreases at periods in excess of a second or so, since electrolytic timing capacitors must then be used and these inevitably suffer from poor tolerance and very high and unpredictable leakage currents.

An easy solution to this problem is shown in *Figure 5.38*, where the 555 is used as a gated astable that generates clock signals with fairly short cyclic periods (using non-electrolytic timing elements) and these periods are effectively expanded by a factor of 1000 via the digital divider stage, which also controls the 555 gating. Normally, the divider's output is low and the 555 is gated off. Timer action is initiated by briefly closing S_1; this action sets the counter's internal registers to zero and drives its output high, thus gating the 555 on and enabling it to generate astable clock signals. On the arrival of the 1000th clock signal the counter's output switches low and gates the 555 off again, completing the operating cycle. Thus, in *Figure 5.38*, the 555 astable actually generates 9 ms to 100 ms periods, but these are expanded to give accurate outputs in the range 9 to 100 s via the × 1000 divider stage.

Several manufacturers produce dedicated timer ICs that operate in the basic way shown in *Figure 5.38* These come in two basic types: timers that have a single fixed division ratio are known as **precision long-period timers**, and the best known of these is the ZN1034; timers that have a division ratio that can be varied via external electronic signals are known as **programmable long-period timers**, and the best known of these is the 2240 IC.

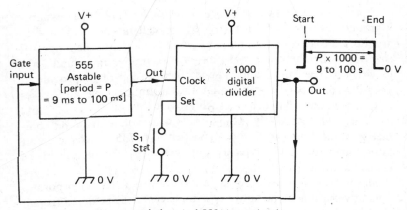

Figure 5.38 *Basic expanded-period 555 timer circuit*

ZN1034 timer circuits

The ZN1034 is a long-period timer that gives precision intervals from 50 ms to 1 h from a single IC, or up to 6 months via a cascaded pair of ICs; the ZN1034 is housed in a 14-pin DIL package with the outline and pin notations shown in *Figure 5.39*, and is intended for direct operation from 5 V TTL supplies; each ZN1034 houses a dual-output twelve-stage binary divider that can be clocked via either an external signal or via an internal precision oscillator that is powered from an internally regulated supply and uses a single C–R timing network; in either case, the divider changes its output states after 4095 complete input cycles.

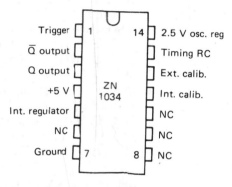

Figure 5.39 *Outline and pin rotations of the ZN1034 precision timer IC*

Figure 5.40 shows the basic timer circuit for the ZN1034. The supply is grounded via pin 7, +5 V are fed to pin 4, calibration pins 11 and 12 are shorted together, and timing component R (value in the range 4k7 to 4M7) is wired between pins 13 and 14 and C (minimum value 3.3 nF) is wired between pin 3 and ground. Brief closing of S_1 provides a negative-going edge on pin 1 that starts a timing cycle and drives output pins 2 and 3 (which can each provide drive currents of 25 mA) low and high respectively. The outputs change state again on completion of the timing cycle, which has a period of $2735 \times C.R$; the circuit table shows four sets of typical component values for periods in the range 1 s to 1 h.

Figure 5.41 shows a three-range ZN1034 timer that incorporates several variations on the above design. The circuit is powered from a 12 V supply by using series resistor R_S to generate a 5 V supply via the

Timer IC generator circuits 119

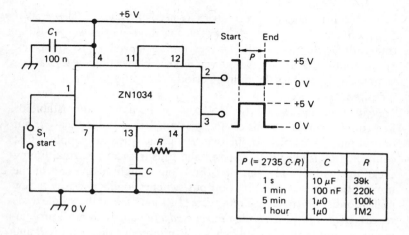

Figure 5.40 Basic ZN1034 timer circuit and typical component values

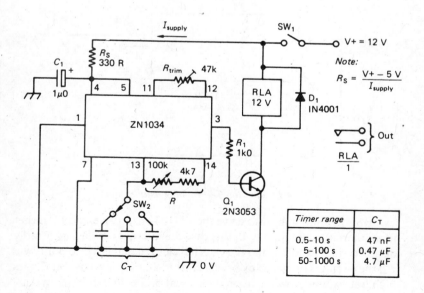

Figure 5.41 Three-range ZN1034 timer with relay output

IC's pin 5 internal shunt regulator (R_S needs a value of $(V+-5V)/I_{supply}$), and a 12 V output relay is driven via inverting/power-boosting transistor Q_1. Supply switch-on triggering is achieved by grounding pin 1, and multi-range timing is achieved by switch-selecting the C_T values; trimmer R_{trim} enables the timing periods to be adjusted by up to 25% if desired.

Figure 5.42 shows how the ZN1034 can be used in the astable or free-running mode. In this case trigger pin 1 is grounded via C_1, but output pin 3 is coupled to pin 1 via R_1. Thus, when the supply is first connected C_1 is effectively shorted and thus triggers a normal timing cycle, but within a few ms pin 3 pulls pin 1 high; at the end of the normal timing cycle the pin 3 output switches low, and within a few milliseconds this transition retriggers pin 1 again, and the whole cycle repeats over. The process then repeats *ad infinitum*, with pin 3 producing an output that briefly switches low at intervals of $2735 \times C.R$ and pin 2 giving an inversion of this action.

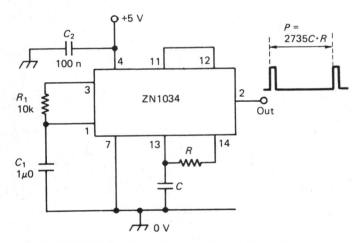

Figure 5.42 *ZN1034 wired in the free-running astable mode*

Figure 5.43 shows how to drive the ZN1034 from an external (rather than internal) clock signal. In this case the internal oscillator is simply disabled via R_2 and C_2', and the external clock signal is fed to pin 12 via R_1; the circuit is triggered by briefly closing S_1, in which case pin 3 goes high but switches low again on the arrival of the 4095th input cycle.

Finally, to complete this look at the ZN1034, *Figure 5.44* shows how the above two circuits can be cascaded to make an ultra-long-period

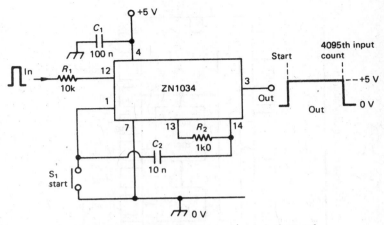

Figure 5.43 *ZN1034 timer driven via an external clock signal*

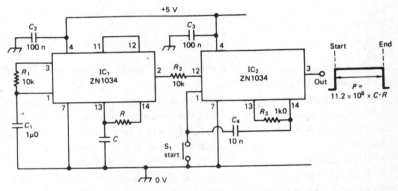

Figure 5.44 *Cascaded ultra long period ZN1034 timer circuit*

timer that can give delay periods up to several months. In this case IC$_1$ is used as a long-period astable (as in *Figure 5.42*) and IC$_2$ is used as an externally-clocked divide-by-4095 unit (as in *Figure 5.43*) the design thus gives a total division ratio of 16,769,025 and gives a time delay of $11.2 \times 10^6 \times C.R$.

2240 timer circuits

The μA 2240 programmable long-period timer IC is a special-purpose device suitable for use in a very limited number of timing applications.

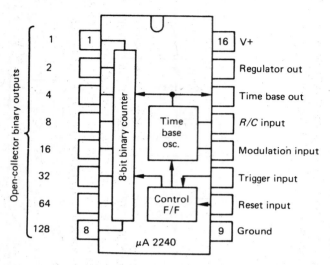

Figure 5.45 *Outline, pin notations, and block diagram of the μA 2240 programmable timer IC*

Figure 5.45 shows the IC's pin notations and block diagram, and Figure 5.46 shows it used in a practical timer circuit.

The IC houses a timebase oscillator, a logic control network, and an 8-bit binary counter with open-collector outputs that give division ratios of 1, 2, 4, 8, 16, 32, 64, and 128 between pins 1 to 8 respectively. The oscillator is controlled via an external R–C timing network (in which R can vary from 1k0 to 10M, and C can vary from 10 nF to 1000 μF) and generates a cyclic period of $C.R$ seconds. When the IC is wired as shown in Figure 5.46 the circuit operates as follows:

The timing operation is initiated by feeding a positive-going trigger pulse to pin 11; this pulse starts the timebase oscillator and sets all divider outputs (and the output of the circuit) to the low state. The IC's eight open-collector binary outputs can be connected in parallel in any desired combination via programming switches S_1 to S_8, in which case R_2 acts as a common collector load that gives a wired-AND logic action in which the output goes high only when all ANDed outputs are high, and this occurs only on the arrival of the nth clock pulse, where n is the divisor equal to the sum of all ANDed binary divider ratios. Thus if switches S_3, S_6 and S_7 are closed, n equals $4 + 32 + 64 = 100$, so the circuit's output switches high on the arrival of the hundreth clock pulse. As the output goes high it feeds a reset signal to pin 10 via R_3,

Timer IC generator circuits 123

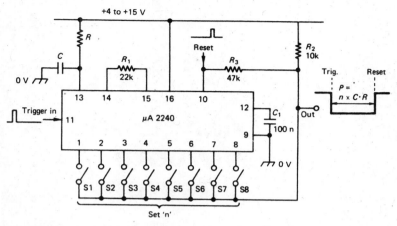

Figure 5.46 µA 2240 programmable timer circuit

and this turns the timebase oscillator off and sets all eight divider outputs high, thus completing the timing operation.

Thus, the above circuit generates a timing period of $n \times R.C$, where n can be programmed to any full-number value in the range 1 to 255 via switches S_1 to S_8.

6 Triangle and sawtooth generators

Triangle and sawtooth waveforms have many uses in electronics, and can be generated in a variety of ways. **Triangle waveforms** are particularly useful for checking crossover distortion in audio amplifiers or for modulating audio signals to produce special sound effects, etc. **Sawtooth waveforms** can be used as timebases for oscilloscopes and wobbulators, etc. A variety of practical waveform generators of these types are described in this chapter.

UJT sawtooth generators

The **unijunction transistor** (UJT) is particularly useful for generating sawtooth waveforms. *Figure 6.1* shows a simple wide-range UJT oscillator circuit that gives a non-linear sawtooth output. Here, in each operating cycle, C_1 first charges exponentially towards the positive supply rail via RV_1–R_1 until it reaches the *peak point* or firing voltage of the UJT, at which point the UJT turns on and rapidly discharges C_1 until its discharge current falls to the UJT's *valley point* or unlatching value, at which point the UJT turns off and C_1 starts to recharge again, and the sequence repeats *ad infinitum*.

Thus, a free-running non-linear (exponential) sawtooth is generated across C_1 and can be fed to external circuits via buffer transistors Q_2–Q_3 and RV_2. With the C_1 value shown the frequency is variable from 25 Hz to 3 kHz via RV_1; it can be varied from below 0.1 Hz to above 100 kHz by using alternative C_1 values.

The UJT circuit can be made to generate a linear sawtooth

Triangle and sawtooth generators 125

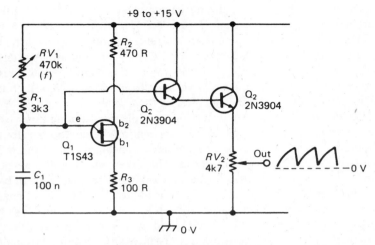

Figure 6.1 *Simple wide-range (25 Hz–3 kHz) non-linear sawtooth generator*

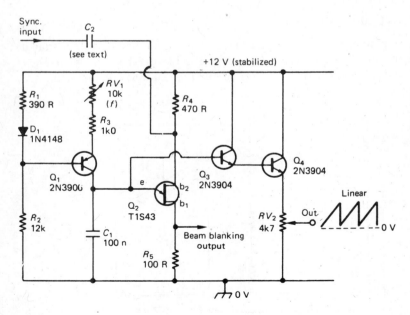

Figure 6.2 *Linear sawtooth generator can be used as an oscilloscope timebase generator*

waveform by charging its timing capacitor (C_1) from a constant-current source, as shown in *Figure 6.2*. Here, Q_1 is used as a temperature-compensated constant-current generator, with current variable from 35 µA to 390 µA via RV_1. C_1's linear sawtooth is externally available via Q_3–Q_4 and RV_2. With the component values shown the operating frequency is variable from 60 Hz to 700 Hz, but other ranges can be obtained by altering the C_1 value.

The *Figure 6.2* circuit can be used as a simple free-running timebase generator for an oscilloscope. In this application the sawtooth output should be fed to the oscilloscope's **external timebase** socket, and the positive flyback pulses from R_5 can be taken via a high-voltage blocking capacitor and used for beam blanking. The sawtooth can be synchronized to an external signal that is fed to Q_2 via C_2; this signal (which must have a peak amplitude between 200 mV and 1V0) modulates the triggering points of Q_2 and thus synchronizes the oscillator and input signals. Note that C_2 must have an impedance less than 470 Ω at the sync signal frequency; if the sync signal is rectangular, with short rise and fall times, C_2 can simply be given a value of 470 pF.

Op-amp circuits

Op-amps can readily be made to produce good low-frequency triangle or ramp waveforms. One easy way of achieving this is shown in *Figure 6.3*, where IC_1 is wired as a simple relaxation oscillator of the type

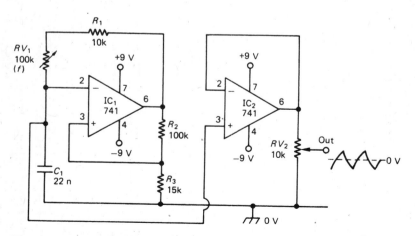

Figure 6.3 *800 Hz–8 kHz non-linear triangle waveform generator*

shown in *Figure 3.9* (which generates a symmetrical but slightly non-linear triangle waveform across C_1 and a square wave at pin 6) and IC_2 is used as a unity-gain voltage follower that acts as a buffer between the C_1 waveform and the external load.

The basic action of the *Figure 6.3* circuit is such that C_1 first charges in a positive direction via R_1–RV_1 and IC_1's output until a certain trigger voltage is reached, at which point the op-amp output changes state and C_1 discharges towards a negative value via R_1–RV_1 and the op-amp output until a second firing voltage is reached, at which point the op-amp output switches back to its original state and the sequence repeats *ad infinitum*. A reasonably linear triangle waveform, with a peak-to-peak amplitude of 1.7 V, is developed across C_1, and its frequency is variable from 800 Hz to 8 kHz via RV_1.

The *Figure 6.3* circuit generates a symmetrical triangle waveform, since C_1 charges and discharges via the same resistance network. *Figure 6.4* shows how the design can be modified to act as a 1 kHz variable-symmetry (variable slope) ramp generator. Here, in each operating cycle, C_1 alternately charges via R_1–D_1 and the left-hand side of RV_1 and discharges via R_1–D_2 and the right-hand side of RV_1. Both resistance paths are variable over an 11:1 range; the rise:fall period ratios of the waveform are thus fully variable from 1:11 to 11:1 via RV_1. The operating frequency is variable over a limited range via RV_2, and has a nominal value of 1 kHz.

The triangle/ramp generator circuits of *Figures 6.3* and *6.4* each generate an output that is slightly non-linear. *Figure 6.5* shows the basic circuit of an op-amp function generator that gives a truly linear triangle output waveform. Here, IC_1 is wired as an integrator, driven from the output of IC_2, and IC_2 is wired as a differential voltage comparator, driven from the output of IC_1 via potential divider R_2–R_3, which is connected between the outputs of IC_1 and IC_2. The square-wave output of IC_2 switches alternately between positive and negative saturation. The circuit functions as follows:

Suppose initially that IC_1's output is positive and IC_2's output has just switched to positive saturation. The inverting input of IC_1 is a virtual earth point, so a current (i) of $+V_{sat}/R_1$ flows into R_1, causing IC_1's output to start swinging down linearly at a rate of i/C_1 volts per second. This output is fed, via the R_2–R_3 divider, to IC_2's non-inverting input, which has its inverting terminal referenced directly to ground.

Consequently, the output of IC_1 swings linearly to a negative value until the R_2–R_3 junction voltage falls to zero, at which point IC_2 enters a regenerative switching phase in which its output abruptly switches to

128 Timer/Generator Circuits Manual

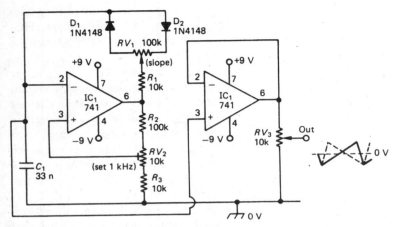

Figure 6.4 *1 kHz variable-slope ramp generator*

negative saturation. This reverses the inputs of IC_1 and IC_2, so IC_1 output starts to rise linearly, until it reaches a positive value at which the R_2–R_3 junction voltage reaches the zero volts reference value, initiating another switching action. The whole process then repeats *ad infinitum*.

Important points to note about the *Figure 6.5* circuit are that the peak-to-peak amplitude of the linear triangle waveform is controlled by the R_2–R_3 ratio, and that the operating frequency can be altered by changing either the ratios of R_2–R_3, the values of R_1 or C_1, or by feeding R_1 from a potential divider connected to the output of IC_2 (rather than directly from IC_2 output). *Figure 6.6* shows the practical circuit of a variable-frequency triangle-wave generator using the latter technique.

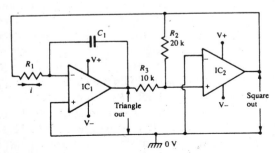

Figure 6.5 *Basic linear triangle wave generator*

Triangle and sawtooth generators 129

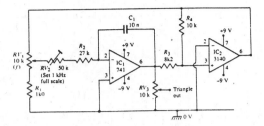

Figure 6.6 *100 Hz–1 kHz linear triangle wave generator*

In *Figure 6.6*, the input current to C_1 (obtained from RV_2–R_2) can be varied over a 10:1 range via RV_1, enabling the frequency to be varied from 100 Hz to 1 kHz; RV_2 enables the full-scale frequency to be set to precisely 1 kHz. The amplitude of the linear triangle output waveform is fully variable via RV_3.

The *Figure 6.6* circuit generates symmetrical output waveforms, since C_1 alternately charges and discharges at equal current values (determined by RV_2–R_2, etc.). *Figure 6.7* shows how the above circuit can be modified to make a variable-symmetry ramp generator, in which the slope of the ramp and the M/S ratio of the rectangle is variable via RV_2. C_1 alternately charges via R_2–D_1 and the upper half of RV_2, and discharges via R_2–D_2 and the lower half of RV_2.

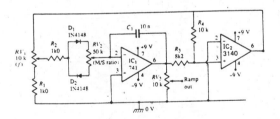

Figure 6.7 *100 Hz–1 kHz variable-symmetry ramp waveform generator*

555 generator circuits

A 555 timer IC can be used to generate a symmetrical but slightly non-linear triangle waveform by wiring it as a conventional astable multivibrator (like *Figure 5.17*) and taking the triangle waveform from across its main timing capacitor via a buffer transistor stage. *Figure 6.8* shows the practical version of such a circuit, in which the frequency is

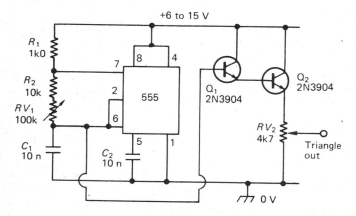

Figure 6.8 *650 Hz–7.2 kHz non-linear triangle wave generator*

variable from 650 Hz to 7.2 kHz via RV_1 and the waveform amplitude is variable via RV_2.

One of the most important applications of the 555 is as a triggered sawtooth generator, and *Figure 6.9* shows a version of the circuit that generates a non-linear (exponential) output. Here, the 555 is wired as a modified monostable multivibrator that is triggered by an external

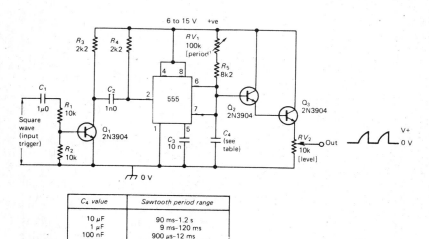

C_4 value	Sawtooth period range
10 µF	90 ms–1.2 s
1 µF	9 ms–120 ms
100 nF	900 µs–12 ms
10 nF	90 µs–1.2 ms
1 nF	9 µs–120 µs

Figure 6.9 *Triggered sawtooth generator*

square wave via Q_1 and C_2 etc., and in which the sawtooth output waveform is taken from across the C_4 timing capacitor via buffer transistors Q_2–Q_3 and via RV_2.

The *Figure 6.9* circuit action is such that the C_4 voltage is normally zero, but each time the 555 is triggered C_4 charges exponentially via R_5 and RV_1 to 2/3 V_{cc}, at which point the monostable period ends and the C_4 voltage switches back to zero. The sawtooth's period can be varied over a decade range via RV_1, and the range can be varied between 9 μs and 1.2 s by using the C_4 values shown in the table; the maximum usable repetition frequency is about 100 kHz.

The basic *Figure 6.9* circuit can be made to produce a triggered linear sawtooth waveform by charging C_4 via a constant-current generator, as shown in *Figure 6.10*. Here, Q_1 is used as the constant-current generator, and the output waveform is taken from across C_4 via Q_2 and RV_2.

When the capacitor is charged via a constant current generator its voltage rises linearly at a rate of I/C volts per second, where I is the charge current in amps and C is the capacitance in farads. Using more practical quantities, the rate of voltage rise can also be expressed as milliamperes/microfarads volts per millisecond. Note that rise rate can be increased by either increasing the charge current or decreasing the capacitance value.

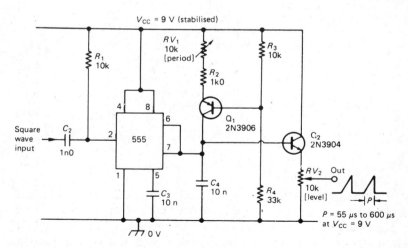

Figure 6.10 *Triggered linear sawtooth generator*

In the *Figure 6.10* circuit the charging current can be varied from 90 µA to 1 mA via RV_1, thus giving rates-of-rise of 9 V/ms to 100 V/ms respectively on the 10 nF timing capacitor. Now, each monostable cycle of the 555 ends at the point where the C_4 voltage reaches $2/3\ V_{cc}$, so, assuming that a 9 V supply is used (giving an end value of 6 V), it can be seen that the circuit's sawtooth cycles have periods variable from 666 µs (= 6/9 ms) to 60 µs (= 6/100 ms) respectively. Periods can be increased beyond these values by increasing the C_4 value, or vice versa. Note that the circuit's supply rail voltage must be stabilized, to give stable timing periods.

Figure 6.11 shows the above circuit modified for use as an oscilloscope timebase generator. The 555 is triggered by square waves derived from external waveforms via a suitable trigger selector circuit, and the ramp output waveform is fed to the scope's X plates via a suitable amplifier stage; the pin 3 output of the 555 provides bright-up pulses to the Z axis of the scope tube during the ramp period, ensuring that the tube is blanked when the timebase is inactive.

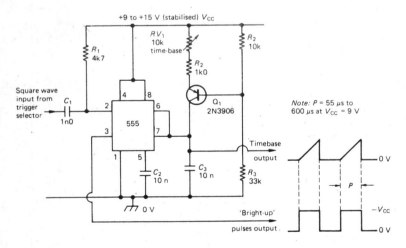

Figure 6.11   *Oscilloscope timebase generator circuit*

Note that the minimum useful ramp period that can be obtained from *Figure 6.11* (using a 1n0 capacitor in the C_3 position) is about 5 µs which, when expanded to give full deflection on a ten-division scope screen, gives a maximum timebase speed of 0.5 µs per division. The circuit gives excellent signal synchronization at trigger frequencies

Triangle and sawtooth generators 133

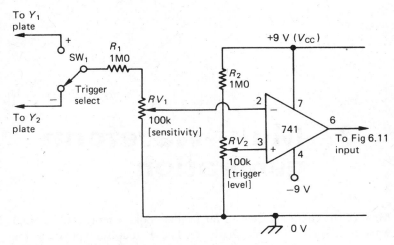

Figure 6.12 *Trigger selection circuit for use with Figure 6.11*

up to about 150 kHz; at higher frequencies the input trigger signals should be reduced via a single- or multi-decade frequency divider. Using this technique, the timebase can be used to view input signals up to many megahertz.

Finally, *Figure 6.12* shows a simple but effective trigger selector circuit that can be used with the above circuit. It comprises a 741 (or better) op-amp voltage comparator circuit, which has a reference voltage fed to its non-inverting input terminal via the RV_2 **trigger level** control pot, and the signal voltage fed to its inverting terminal via SW_1–R_1 and the RV_1 **sensitivity pot**. SW_1 enables either in-phase or antiphase input signals to be selected from the Y driving amplifier, thus enabling either positive or negative trigger modes to be selected. The output of the *Figure 6.12* circuit is coupled directly to the C_1 input of *Figure 6.11*.

7 Multi-waveform generation

Most of the circuits shown so far in this book generate only a single waveform, such as sine or square, etc. The most useful types of generator, however, are those that can produce several waveform types, either simultaneously or alternatively (via switched *mode selectors*). Some dedicated waveform synthesiser ICs do offer these facilities, and two such devices (the XR-2206 and the ICL8038) are described in Chapter 8. These ICs are rather expensive, however, so this chapter looks at some easy and inexpensive ways of generating multiple waveforms; a total of ten different circuits are shown.

Op-amp circuits

Figure 7.1 shows an op-amp circuit that simultaneously generates both sine and square waves and spans 150 Hz to 1.5 kHz in a single frequency range. Here, IC_1 is wired as a lamp-stabilized Wien-bridge oscillator of the type shown in *Figure 2.5*, and its direct sine-wave output is fed to the input of IC_2, which is wired as a simple voltage comparator and has its reference terminal tied to the zero volts rail. The action of IC_2 is such that its output switches from positive saturation to negative saturation, or vice versa, each time the sine-wave input swings through the zero-volts level, thus converting the sine-wave input into a useful square-wave output.

The *Figure 7.1* circuit is initially set up by adjusting RV_2 to give 2 V r.m.s sine-wave output at the maximum setting of RV_3. Once set, the circuit generates a sine wave with a THD content of about 0.1%. The square-wave output is available simultaneously, and has an amplitude that is variable via RV_4 and has typical rise and fall times of about 1 μs at 1 kHz.

Multi-waveform generation

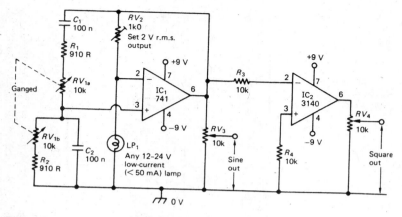

Figure 7.1 150 Hz–1.5 kHz sine/square generator

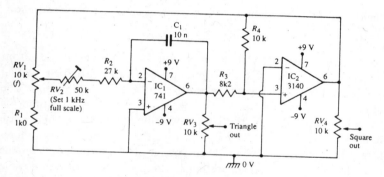

Figure 7.2 100 Hz–1 kHz triangle/square generator

Figure 7.2 shows an op-amp circuit that simultaneously generates symmetrical linear-triangle and square waves and spans 100 Hz to 1 kHz in a single frequency range. This is a simple development of the linear triangle generator circuit shown in *Figure 6.6*, which develops a linear triangle output on the slider of RV_3 but also generates a synchronous square wave on the output of IC_2; in *Figure 7.2* this square wave is made available at a variable amplitude level on the slider of RV_4. The square-wave output gives typical rise and fall times of less than 1 μs.

The linear output triangle of the *Figure 7.2* circuit can be converted into a good sine wave or into a variable mark/space ratio rectangular wave with the aid of simple adaptors. *Figure 7.3* shows a practical variable M/S ratio adaptor. Here, the op-amp is wired as a simple

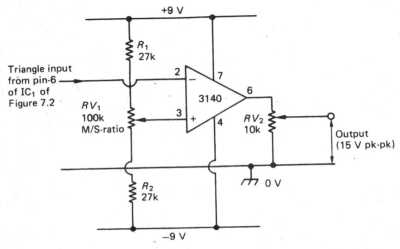

Figure 7.3 *Variable M/S ratio rectangle wave adaptor for use with the Figure 7.2 circuit*

voltage comparator and has one input applied from the triangle output of *Figure 7.2* and has its other input applied from a variable potential divider that is wired between the positive and negative supply lines. The op-amp switches into positive or negative saturation each time the triangle voltage goes more than a few millivolts below or above the reference voltage set by RV_1. By adjusting the reference voltage, therefore, the op-amp can be made to change state at any point on the triangle waveform, thus generating a variable M/S ratio square-wave output. Note that the *Figure 7.2* and *7.3* circuits enable both frequency and M/S ratio to be varied with zero interaction between the two controls.

Figure 7.4 shows a circuit that can be used to convert the linear triangle waveform of *Figure 7.2* into a fairly good sine wave. Here, the triangle waveform is fed into a resistor-diode matrix via adjustable potential divider RV_1–R_1. The matrix converts the triangle waveform into a simulated sine wave by automatically reducing the slope of the triangle in a series of steps as the triangle amplitude increases. The resulting waveform is fed into a ×2.2 non-inverting d.c. amplifier, and is finally available with a maximum peak-to-peak amplitude of 14 V across variable output control RV_2. The diagram shows that the final output waveform can be represented by a series of straight lines, there being four of these in each quarter cycle. The simulated sine wave

Multi-waveform generation 137

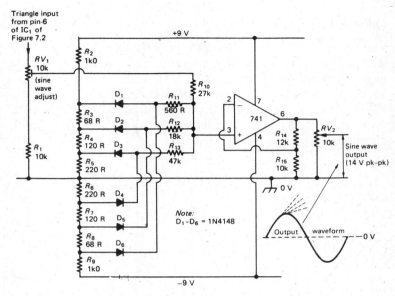

Figure 7.4 *Sine wave adaptor for use with the Figure 7.2 circuit*

typically contains less than 2% THD. RV_1 should be adjusted to give the best sine wave shape when the converter is initially connected to the *Figure 7.2* circuit.

Finally, to complete this look at op-amp-based multi-waveform generators, *Figure 7.5* shows how the variable-slope linear ramp generator of *Figure 6.7* can be modified so that it also produces a simultaneous variable-amplitude variable M/S ratio rectangular output waveform by simply wiring RV_4 between the output of IC_2 and ground.

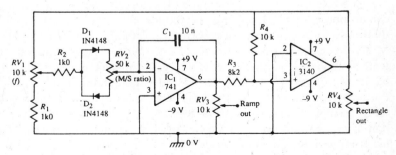

Figure 7.5 *100 Hz–1 kHz ramp/rectangle generator with variable slope and M/S ratio*

138 Timer/Generator Circuits Manual

Unijunction-based circuits

Unijunction (UJT) circuits can easily be made to generate both sawtooth and pulse or rectangular waveforms. *Figure 7.6*, for example, shows a free-running 25 Hz to 3 kHz UJT oscillator that generates a non-linear output on RV_2 and a 30 µs positive pulse across R_3 and a 30 µs negative pulse across R_2. The waveform period is determined by the values of R_1–RV_1 and C_1, and the pulse widths are determined by the values of C_1 and R_3.

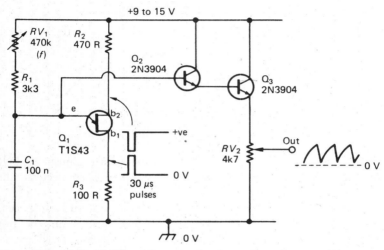

Figure 7.6 *25 Hz–3 kHz non-linear sawtooth generator also gives 30 µs output pulses*

Figure 7.7 shows how a free-running UJT oscillator can be used in conjunction with a 741 op-amp to produce either a non-linear sawtooth waveform or a rectangular waveform with an infinitely variable M/S ratio, depending on the setting of SW_1. When SW_1 is in the **sawtooth position** the op-amp acts as a unity-gain voltage follower, with its input fed from the C_1 waveform, and under this condition the circuit thus generates a sawtooth output via RV_2. When SW_1 is in the **rectangle position** the op-amp acts as a simple voltage comparator, with its non-inverting input fed from the C_1 sawtooth and its inverting input fed from variable potential divider R_4–RV_3–R_5, which is connected across the circuit's supply lines; the comparator's action is such that its output switches rapidly from positive to negative saturation, or vice

Multi-waveform generation 139

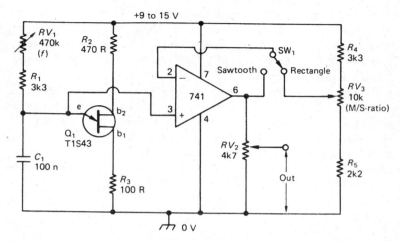

Figure 7.7 *25 Hz–3 kHz generator produces a non-linear sawtooth or a rectangle waveform with infinitely variable M/S ratio*

versa, as the instantaneous sawtooth input voltage passes through the voltage level set on the slider of RV_3. This transition occurs twice in each sawtooth cycle, and the circuit thus generates a rectangular output waveform across RV_2 under this condition. The M/S ratio of this waveform is infinitely variable from 0:1 to 1:0 via RV_3.

The *Figure 7.7* circuit can be modified so that it generates a linear sawtooth waveform by simply using a constant-current generator to charge timing capacitor C_1, as shown in the circuit of *Figure 7.8*, which

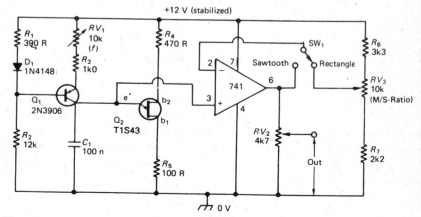

Figure 7.8 *60 Hz–700 Hz generator produces a linear sawtooth or a variable M/S ratio rectangle output*

spans the 60 Hz to 700 Hz frequency range with the C_1 value shown. Note that the operating frequencies of these two circuits can be increased by simply reducing the value of C_1, but that if frequencies greater than 10 kHz are required a 3140 or similar fast op-amp should be used in place of the 741.

Finally, to complete this look at UJT-based designs, *Figure 7.9* shows how the *Figure 7.7* circuit can be modified so that it provides simultaneous sawtooth and variable M/S ratio rectangular waveforms. In this case the op-amp is permanently wired in the voltage comparator mode, with its input fed from the C_1 sawtooth, and provides a rectangular output across RV_3, and C_1's sawtooth waveform is made externally available via RV_2 and buffer transistors Q_2–Q_3.

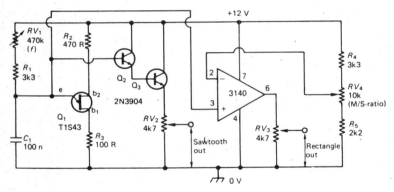

Figure 7.9 *25 Hz–3 kHz sawtooth/rectangle generator*

A triggered pulse/sawtooth generator

To complete this chapter, *Figure 7.10* shows how a 555 timer IC can be used as a triggered generator that provides a positive pulse output on RV_1 and a simultaneous linear sawtooth on RV_3. The two waveforms have identical timing periods, and span the range 55 µs to 600 µs with the C_3 value shown. Note that this circuit is a simple modification of the *Figure 6.10* design, and a more complete description of its operation is given in Chapter 6.

Multi-waveform generation 141

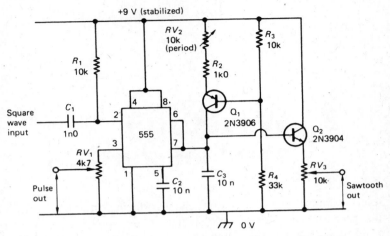

Figure 7.10 *Triggered 55 µs to 600 µs pulse and sawtooth generator*

8 Waveform synthesizer ICs

Several manufacturers produce special waveform synthesizer or function generator ICs. The most useful of these can generate several waveforms, including good sine, square and triangle types, and their design is based on an integral voltage-controlled oscillator (VCO) that enables the operating frequency to be easily controlled or modulated. The best known of these devices are the XR-2206, from Exar Integrated Systems Inc., and the 8038, from Intersil. This chapter deals with both of these ICs.

XR-2206 basics

The XR-2206 is a high-quality device that can generate good sine, square and triangle waveforms at frequencies from a fraction of a hertz to several hundred kilohertz, using a minimum of external circuitry. Its frequency can be swept over a 2000:1 range via an external voltage or resistance, and the IC incorporates facilities for imposing AM, FM, phase-shift, and FSK modulation on various of its output waveforms. The device has excellent frequency stability (typically 20 p.p.m./°C for thermal changes and 0.01%/V for supply voltage changes) and can be powered from either a single-ended supply in the range 10 to 26 V or a split supply in the range ± 5 to ± 13 V. Sine-wave outputs have a typical THD content of 2.5%, but this can easily be reduced to about 0.5% via two trimmer controls. The sine wave has a maximum amplitude of 2 V r.m.s. and an output impedance of 600 Ω.

Figure 8.1 shows the outline, pin notations, and functional block diagram of the XR-2206, which is housed in a 16-pin DIL package. The VCO and its current switches form the heart of the IC; to operate,

Waveform synthesizer ICs 143

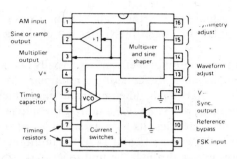

Figure 8.1 *Functional block diagram and pin notations of the XR-2206 function generator IC*

the VCO must be connected to an external capacitor (which is wired between pins 5 and 6 and can have any value from 1n0 to 100 μ) and a resistor (which can have any value between 1k0 to 2M0 and can be wired between either pin 7 *or* pin 8 and the negative supply rail). The current switches simply select the resistor timing pin that is to be used. Normally, with pin 9 floating, timing pin 7 is selected, but if pin 9 is pulled close to the negative supply rail voltage pin 8 is selected. Pin 9 thus enables alternate timing resistors to be selected at will, to give either frequency-shift keyed (FSK) modulation, or to facilitate the generation of non-symmetrical square and ramp waveforms.

Suppose, then, that a timing capacitor (C) is placed across pins 5 and 6, a timing resistor (R) is in place on pin 7, and that pin 9 is floating. In this case the VCO action is such that C first charges linearly at a rate determined by R until a certain firing voltage is reached, at which point the VCO changes state and C starts to discharge linearly at a rate again determined by R until a second firing point is reached, at which point the process starts to repeat. The consequence is that the VCO generates a linear and symmetrical sawtooth waveform that is passed on to pin 2 via the **multiplier and sine shaper block**, and simultaneously generates a symmetrical square wave that is fed to pin 11 via the integral buffer transistor. The VCO's operating frequency is given by:

$$f_o = 1/RC \, \text{Hz}$$

and can be varied via either R or C.

The IC's VCO is actually a current-controlled multivibrator in which the timing current is controlled by the resistors connected to pins 7 or 8, or by voltages or signals that are connected to these pins via suitable current-limiting resistors. This factor makes it possible to externally modulate or sweep the frequency of the generated VCO signals.

The VCO's ramp output waveform is internally connected to a pair of input terminals on a four-quadrant multiplier and sine shaper, which acts like a variable-gain differential amplifier which gives a high-impedance output at pin 3 and (via a unity gain buffer) a 600 Ω ouput impedance at pin 2 and has its gain externally controlled via a bias applied to pin 1. When pins 13 and 14 of the multiplier are open, the ramp waveform passes through the multiplier and appears at pins 2 and 3, but when a resistance of a few hundred ohms is connected between pins 13 and 14 the multiplier exponentially limits the peaks of the ramp waveform and converts it into a sine wave that appears at pins 2 and 3. With suitable adjustment, the sine wave distortion can be reduced to about 0.5%.

The multiplier's gain and phase are linearly controlled by variations in the pin 1 voltage around the half-supply value. The gain (and output) is zero when pin 1 is at half-supply volts, but rises as the voltage is increased. When the voltage is reduced below the half-supply value the gain (and output level) again increase, but the signal phase is reversed. These characteristics enable the sine/ramp waveforms to be externally amplitude modulated or phased-shift keyed via signals applied to pin 1.

Note that the multiplier's high impedance output appears at pin 3, but that this point is also internally connected to the input of the buffer that gives the low-impedance pin 2 output. Consequently, the buffer's input signal (and hence the pin 2 output) can be effectively varied by potential divider action by simply wiring a variable resistor between pin 3 and a virtual ground point. This facility can be used to provide simple amplitude control of the pin 2 output signal, or to facilitate gate keying or pulsing of the output signal.

A final point to note is that the d.c. level of the pin 2 output is approximately the same (within a few hundred millivolts) as the d.c. bias voltage of pin 3. Thus, d.c. level shifting can be applied to the pin 2 output by applying a suitable bias to pin 3. In most applications pin 3 is biased at half-supply volts. In split-supply circuits this means that the pin 2 output signal swings about the zero volts (common) line.

Sine-wave generation

The XR-2206 is very easy to use in basic waveform generator applications. *Figure 8.2* shows how to wire it as a simple wide-range sine-wave generator that is powered from a single-ended 12 to 18 V supply. R_3–RV_1 forms the main timing resistance; it is connected between pins 7 and 12 (ground) and is automatically selected because pin 9 (FSK input) is left floating. The sine-wave frequency is variable over a decade range via RV_1 with any given value of C_3, as indicated. The circuit generates a sine-wave output at pin 2, since a 220 Ω resistor is wired between pins 13 and 14. Typically, the sine-wave distortion is less than 2.5% with this simple connection; the sine wave output amplitude is fully variable via RV_2.

In *Figure 8.2* the pin 3 voltage is biased at half-supply volts via decoupled divider R_1–R_2, so the pin 2 output is also biased near half-supply volts. RV_3 enables the pin 2 sine wave amplitude to be pre-set to

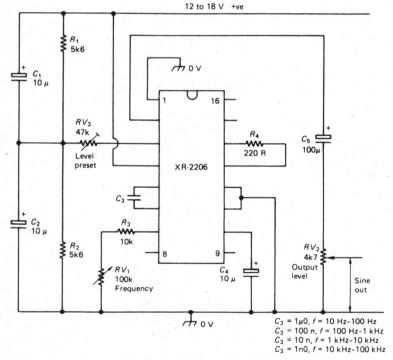

Figure 8.2 *Simple single-supply, wide-range sine-wave generator*

a value at which distortion (due to clipping) is minimal. To set RV_3, first disconnect R_4 (so that a triangle output is obtained), then adjust RV_3 so that no triangle clipping occurs, then re-connect R_4 and check that a decent sine wave is available. If desired, sine-wave distortion can be reduced below the typical 2.5% value by replacing R_4 with a 470R pre-set and adjusting it for minimum distortion.

Figure 8.3 shows how the above sine-wave generator can be modified for split-supply operation, by replacing all ground connections with negative-rail ones and by taking **level preset** control RV_3 to the common supply (ground) line. This circuit also shows how the sine wave's total harmonic distortion (THD) can be reduced to about 0.5% with the use of presets RV_4 and RV_5. These controls must be adjusted alternately to give the best possible sine-wave output, after first setting RV_3 to give a non-clipped triangle wave output as already described.

In the absence of a distortion factor meter, the simple twin-T 1 kHz filter of *Figure 8.4* can be used in conjunction with a scope or a.c. millivoltmeter to set the above sine-wave generator for minimum dis-

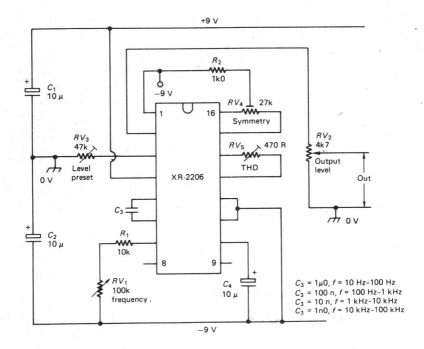

Figure 8.3 *High-performance, split-supply sine-wave generator*

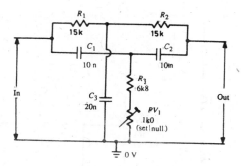

Figure 8.4 *Simple 1 kHz twin-T filter aids setting up for minimum THD*

tortion at 1 kHz. To use the filter, first set the generator output to 1 V r.m.s. at 1 kHz and connect it to the filter input, and connect the filter output to the input of an indicator such as an oscilloscope or millivoltmeter. Next, trim the frequency of the generator and RV_1 of the filter to give minimum output indication, and then finally trim the generator's RV_4 and RV_5 distortion controls to give a minimum reading on the indicator. At final balance, the filter's output distortion corresponds to about 0.1% THD per millivolts r.m.s. of indicated reading, i.e., if the indicator shows a reading of 5 mV r.m.s., the THD of the generator approximates 0.5%.

When using the low-distortion sine wave facility of *Figure 8.3*, note that the pin 3 output signal is similar to that of pin 2 but has lower distortion and a higher output impedance. Also, the pin 3 signal is closely centred on the common or ground line, but the pin 2 signal is offset by a few hundred millivolts. If desired, slight d.c. offset can be applied to pin 3 (to bring the pin 2 output to precisely zero offset value) by using the add-on modification shown in *Figure 8.5*.

Triangle and square generation

The XR-2206 can be made to generate a linear triangle output by using the basic circuits of *Figures 8.2* or *8.3* without the sine-shaping resistors. *Figure 8.6* shows a variable frequency split-supply triangle waveform generator that, when used with a 9 V–0–9 V supply, gives maximum peak-to-peak triangle waveform amplitudes of 12 V before clipping occurs.

The XR-2206 can be made to produce fixed-amplitude square waves at pin 11, either independently or simultaneously with sine or triangle

148 Timer/Generator Circuits Manual

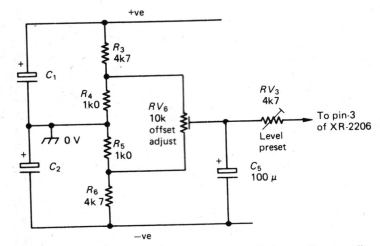

Figure 8.5 *Add-on modification for applying limited d.c. offset or nulling to the output of the Figure 8.3 circuit*

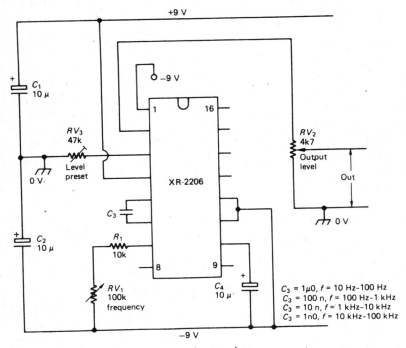

Figure 8.6 Wide-range, split-supply triangle wave generator

Waveform synthesizer ICs 149

waveforms, by simply wiring a 4 k7 load resistor between pins 11 and 4, as shown in the split-supply circuit of *Figure 8.7*. The rise and fall times of the square-wave output signal are typically 250 ns and 50 ns respectively when pin 11 is loaded by 10 pF. *Figure 8.8* shows how a simple CMOS inverter stage can be used as a buffer between pin 11 and the final square-wave output, to give a variable-amplitude output with improved rise and fall times.

The basic sine, triangle and square-wave generating facilities of the XR-2206 can be combined in a variety of ways to make multi-function waveform generators. *Figure 8.9*, for example, shows how various circuits can be combined to make a simple split-supply sine, triangle, or square generator. Here, the fixed-amplitude square wave is taken dir-

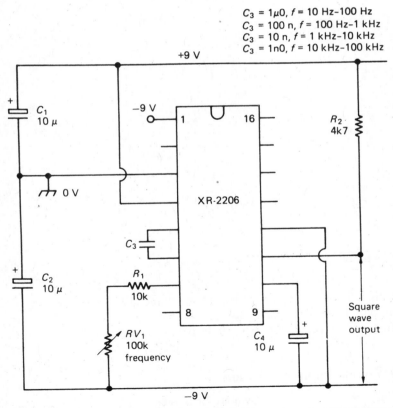

Figure 8.7 *Simple fixed-amplitude variable frequency square-wave generator*

150 Timer/Generator Circuits Manual

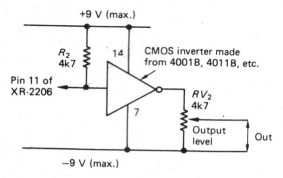

Figure 8.8 *Add-on, variable-amplitude circuit for use with the Figure 8.7 square-wave generator*

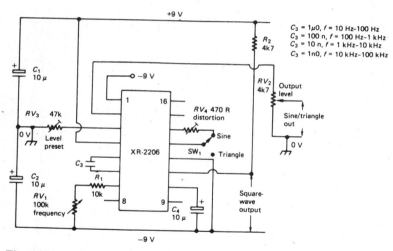

Figure 8.9 *Simple split-supply sine/triangle/square-generator*

ectly from pin 11 and is produced simultaneously with the variable-amplitude sine or triangle waveforms, which are selected via SW_1. The frequency is variable from 10 Hz to 100 kHz via C_3 and RV_1.

Pulse and ramp generation

The XR-2206 circuits shown so far all generate symmetrical output waveforms. The IC can, however, be made to generate non-symmetrical outputs such as ramp, sawtooth and pulse waveforms, by simply shorting the pin 9 FSK terminal to pin 11, as shown in *Figure 8.10*, so that the VCO uses the R_1 set of resistors to time one half of the waveform and uses the R_2 set of components to time the other half of the waveform.

The *Figure 8.10* circuit produces a variable-amplitude variable-slope ramp output waveform from the slider of RV_1, and a simultaneous fixed-amplitude variable M/S ratio square or pulse output from pin 11. The rise and fall or on and off periods of the ramp or square waveforms are independently controlled via R_1 and R_2 and can each be varied over a 100:1 range, giving a total M/S ratio range of 100:1 to 1:100.

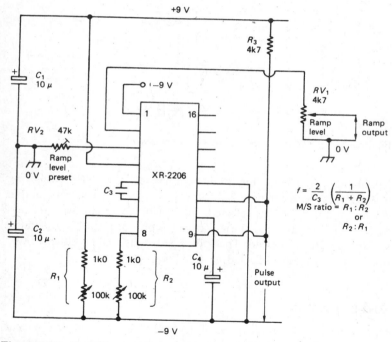

Figure 8.10 *Variable ramp and pulse generator circuit*

Frequency-shift keyed generation

The pin 9 FSK input terminal of the XR-2206 can also be used to provide the waveform generator with frequency-shift keyed (FSK) operation, as shown in *Figure 8.11*. Here, a keying or pulse waveform is

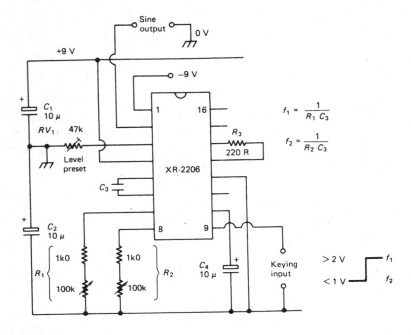

Figure 8.11 *Basic split-supply FSK sine-wave generator*

fed directly to pin 9, and the circuit action is such that when this keying waveform is greater than +2 V relative to the negative supply rail pin 7's R_1 timing resistor is selected, but when the keying waveform is below +1 V (relative to the negative supply rail) pin 8's R_2 timing resistor is selected instead. The FSK signal thus enables either of two operating tones to be selected.

AM generation

One of the many useful features of the XR-2206 is that its pin 2 output can easily be subjected to amplitude modulation (AM) by applying a d.c. bias and a suitable modulating signal to pin 1, as shown in *Figure 8.12*. The pin 2 signal amplitude varies in proportion to that of pin 1 when the pin 1 voltage is within 4 V of the circuit's half-supply value (in split-supply circuits this equals zero volts). When the pin 1 voltage is increased above the half-supply value the pin 2 signal rises in direct proportion. When the pin 1 voltage is reduced below the half-supply

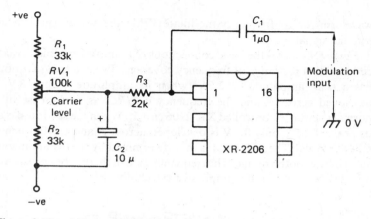

Figure 8.12 *Add-on AM facility (split-supply circuit)*

value the pin 2 signal again rises in direct proportion, but the phase of the output is reversed. This latter phenomenon can be utilized for phase-shift keyed (PSK) and suppressed carrier AM generation.

The pin 1 terminal can also be used to facilitate gate-keying or pulsing of the pin 2 output signal. This can be achieved by biasing pin 1 to near half-supply volts to give zero output at pin 2, and then imposing the gate or pulse signal on pin 1 to raise the pin 2 signal to the desired turn-on amplitude. The available AM dynamic range is about 55 dB.

Frequency sweep and FM generation

The XR-2206's oscillation frequency is proportional to the total timing current (I_T) drawn from pin 7 or 8 and is given by:

$$f = (320 \times I_T)/C$$

where f is in hertz, I_T is in milliamperes, and C is in microfarads. The timing terminals (pins 7 and 8) are low-impedance points and are internally biased at +3 V relative to pin 12. The frequency varies linearly with I_T over the range 1 µA to 3 mA, and can thus be voltage-controlled by applying a 0 to +3 voltage between pin 12 and the timing pin via a suitable resistor, so that the timing current is determined by the resistor value and the *difference* between the internal (+3 V) and external (0–3 V) voltages. This technique can be used to either frequency-sweep the generated signals via an external sawtooth waveform, or to frequency-modulate (FM) the waveforms via an external signal.

Figure 8.13 shows the basic connections of a simple frequency-sweep circuit with a 6:1 range of frequency coverage. The external sawtooth has a peak amplitude of 2.5 V; when its amplitude is zero 3 V is developed across R and the frequency is $1/RC$, as in the case of a normal resistance-controlled XR-2206 circuit. When the sawtooth is at its peak of 2.5 V only 0.5 V is developed across R and the frequency falls to $1/6\,RC$. The frequency is thus determined by the instantaneous value of sawtooth voltage. The frequency can, in theory, be varied over a 2000:1 range using this simple sweep technique.

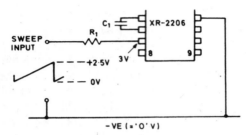

Figure 8.13 *Frequency-sweep circuit giving a 6:1 frequency range*

Figure 8.14 shows the basic way of using the FM (frequency modulation) facility of the XR-2206. Here, in the absence of an FM input signal, the oscillating frequency is determined only by C_1 and R_1, but when an FM input signal is applied its input currents are added to those of R_1 via R_2, so the timing currents (and thus the operating frequency) are modulated by the input signal.

A weakness of the above circuit is that, for a given amplitude of input signal, the percentage deviation or sensitivity of the FM facility varies with the setting of the R_1 frequency control. To complete this

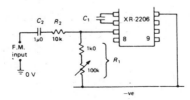

Figure 8.14 *Simple FM facility for the XR-2206*

Waveform synthesizer ICs 155

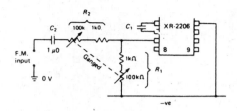

Figure 8.15 *Constant-sensitivity FM facility*

look at XR-2206 applications, the *Figure 8.15* **constant-sensitivity circuit** shows how this snag can be overcome by using a dual-gang pot with one arm connected to the R_1 frequency determining network and the other arm connected to the R_2 FM-sensitivity network, so that the sensitivity is automatically adjusted to track with the frequency setting.

ICL8038 basics

The 8038 waveform generator IC is a medium-quality device that can produce simultaneous sine, square and triangle outputs, with variable duty cycles, at frequencies ranging from a fraction of a hertz to above 100 kHz. Its output signals can be subjected to frequency sweeping and modulation (FM), and it can be powered via either single-ended supplies in the range 10 to 30 V, or split ones in the ±5 to ±15 V range. The 8038 is not as versatile as the XR-2206: it has no facility for generating AM, PSK, or FSK signals, and its sine-wave output distortion is generally higher than that of the XR-2206. The 8038 is, however, a popular device and is widely used as a general-purpose waveform generator.

Figure 8.16 shows the outline and pin notations of the 8038, and *Figure 8.17* shows the basic way of using it as a fixed-frequency triangle, sine, or square waveform generator that is powered via a single-ended supply. Note in this circuit that the pin 7 FM bias terminal is shorted directly to the pin 8 FM sweep input terminal, and in this case the 8038's operating frequency is determined by the values of C, R_A and R_B. These two resistors in fact set the operating values of a pair of internal constant-current generators that alternately charge and discharge the main timing capacitor (C).

The basic action of the *Figure 8.17* circuit is such that, in each operating cycle, C alternately charges linearly at a rate determined by R_A until the C voltage reaches two-thirds of $+V_{cc}$, at which point a switching

156 Timer/Generator Circuits Manual

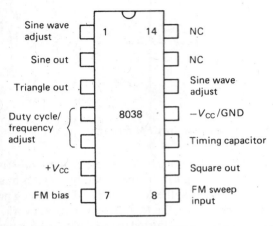

Figure 8.16 *Outline and pin notations of the 8038 waveform generator IC*

action occurs and C starts to discharge linearly at a rate determined by R_B until the C voltage drops to one-third of V_{cc}, at which point another switching action takes place and the whole process starts to repeat. R_A and R_B can have any values in the range 1k0 to 1M0. If these components have equal values (R) the circuit operates at a frequency of $0.3/(RC)$ and generates a symmetrical linear triangle waveform with a

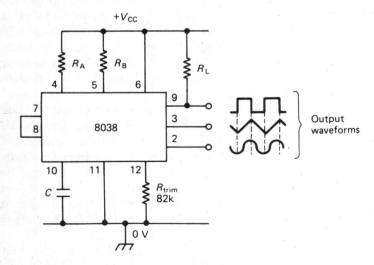

Figure 8.17 *Basic fixed-frequency triangle/sine/square-waveform generator*

peak-to-peak amplitude of $0.33 \times V_{cc}$ on pin 3, and a square wave with a peak-to-peak value of V_{cc} on pin 9 (which is externally loaded by R_L). The basic triangle waveform is also fed to an internal triangle-to-sine converter, which produces a fairly good sine-wave output (with a peak-to-peak amplitude of $0.22 \times V_{cc}$) on pin 2 when R_{trim} is given a value of 82 k as shown.

Sine-wave distortion

Note that the *Figure 8.17* circuit can be made to generate non-symmetrical output waveforms by simply giving R_A and R_B different values. Also note that for best purity of sine-wave generation the circuit must be set to give perfect waveform symmetry, and this can be achieved by modifying the circuit as shown in *Figure 8.18*, which also shows how r_{trim} can be made variable and used to trim the sine wave for minimum distortion. In practice, sine wave THD figures as low as 0.8% can be obtained from the *Figure 8.18* circuit when it is used in fixed-frequency applications below 10 kHz; if desired, distortion can be reduced even further (to about 0.5%) by further modifying the circuit

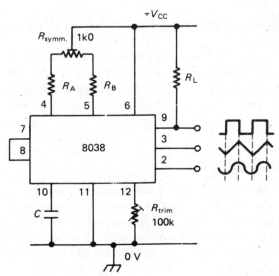

Figure 8.18 *Modified circuit gives perfect symmetry and reduced sine-wave distortion*

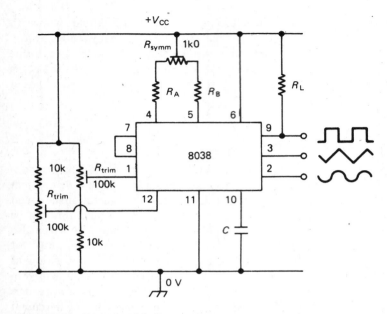

Figure 8.19 *Circuit modified to give minimum sine-wave distortion*

as shown in *Figure 8.19* and trimming all three variable components for best performance.

Before leaving the subject of sine-wave distortion, readers should note that the 8038 rarely maintains perfect symmetry when used in variable frequency applications, and in such cases may produce worst-case THD figures of several per cent. This is the greatest fault of the 8038 IC.

Supplies and buffering

The 8038 is shown using single ended supplies in the circuits of *Figures 8.17* to *8.19*, and in such cases the three output waveforms all swing about (are centred on) the half-supply voltage value. These circuits can be powered via split (dual) supplies, if desired, by simply using the zero rail as the negative supply line; in this case all output waveforms are centred on the zero or ground line of the split supply. Note in all cases that pin 8 of the IC is susceptible to unwanted signal pick-up, and should be decoupled by wiring a 100 nF capacitor between pin 8 and

+ V_{cc}. Also note that the IC consumes a fairly high quiescent current (about 12 mA at 20 V) and tends to run warm.

In most multi-waveform generator applications of the 8038 the user will need to feed the signals to the outside world via some type of buffer circuit, so that the desired waveform can be selected at will and made available at variable amplitude levels. *Figure 8.20* shows how a readily available quad op-amp can be used to make a comprehensive buffer that can be used in split-supply applications and which makes all three output waveforms available at about the same peak amplitude.

FM and sweeping

The operating frequency of the 8038 is a direct function of the DC voltage applied between pin 8 and the IC's positive supply terminal (pin 6). The operating frequency can thus be varied or swept by altering this voltage, or can be modulated by feeding a suitable modulation signal to pin 8.

The easiest way of achieving FM operation (at deviations up to a maximum of ± 10%) is simply to feed the modulating signal to pin 8 via a DC blocking capacitor, as shown in *Figure 8.21*. The external resistor (R) shown between pins 7 and 8 is not essential, but can be used to increase the effective input impedance of pin 7. With R shorted, the input impedance is 8k0; when R has a finite value, the input impedance equals (R + 8k0).

The easiest way of using the 8038 as a manually-controlled variable frequency waveform generator is to wire it as shown in *Figure 8.22*, with pin 8 connected to a variable control voltage taken from RV_1 slider. This voltage is variable from V_{cc} to two-thirds of V_{cc}; the frequency is minimum when the pin 8 voltage equals V_{cc}, and is maximum when it equals (2/3 V_{cc} + 2 V). This simple circuit enables the frequency to be varied over a range of about 40:1.

The maximum sweep range available from the 8038 is at least 1000:1, but to attain this the highest control voltage on pin 8 must exceed that of pin 6 by a few hundred millivolts. This can be achieved by modifying the circuit as shown in *Figure 8.23*, where the IC's pin 6 voltage is reduced to about 600 mV below + V_{cc} by the forward volt drop of diode D_1. Note that, for optimum frequency stability, this circuit's supply voltages must be stabilized.

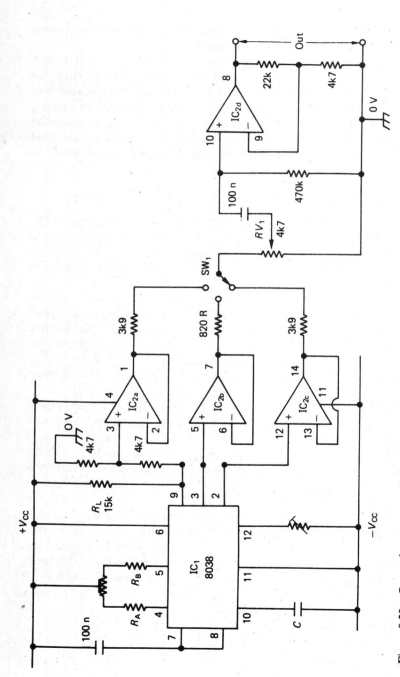

Figure 8.20 *Comprehensive output buffer for use with split-supply 8038 circuits*

Waveform synthesizer ICs 161

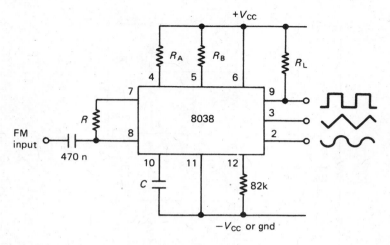

Figure 8.21 Waveform generator with FM facility

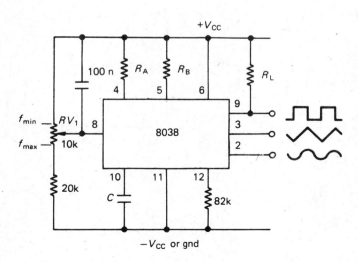

Figure 8.22 Simple variable frequency waveform generator

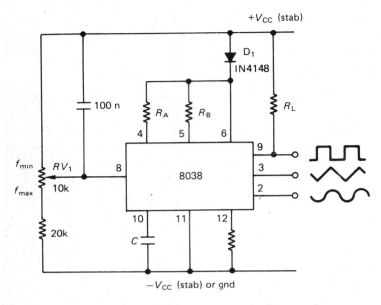

Figure 8.23 Wide-range variable frequency waveform generator

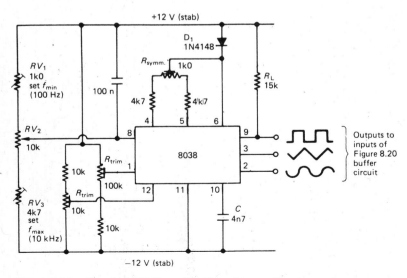

Figure 8.24 Practical 100 Hz to 10 kHz multi-waveform generator

Audio oscillator

To complete this chapter, *Figure 8.24* shows how the 8038 can be used to make a practical variable frequency (100 Hz to 10 kHz) triangle, sine, or square waveform generator by combining the circuits of *Figures 8.19* and *8.23*. To set up this circuit, first set R_{symm} to mid-value and then trim RV_1 and RV_3 so that the generator spans the 100 Hz to 10 kHz frequency range via RV_2, then set RV_2 to give 1 kHz output and trim R_{symm} to give a symmetrical square-wave output. Recheck the frequency span range. Finally, reset 1 kHz and adjust the two R_{trim} controls for minimum sine-wave distortion.

Finally, note that the frequency range of this circuit can be increased (or decreased) by a decade by reducing (or increasing) the value of *C* by a decade.

9 Special waveform generators

In addition to the sine, square, pulse, triangle, and sawtooth generators discussed in earlier chapters, a variety of other special waveform generators are also used in electronics. Examples of these are white noise generators, staircase waveform generators, crystal-controlled oscillators, waveform synthesizers, and special-effects sound generators. This chapter shows a variety of special waveform generator circuits.

Modifying existing waveforms

It was shown in earlier chapters that existing waveforms can often be changed into alternative forms by passing them through simple converter circuitry. Thus, triangle and sawtooth waveforms can be changed into square or rectangle form by passing them through a Schmitt trigger, and triangle waveforms can be converted into sine waves by feeding them through a suitable diode shaping matrix. *Figures 9.1* to *9.5* show a variety of ways of modifying existing waveforms by passing them through simple C–R differentiating and diode clipping networks.

Figure 9.1 shows how the shape and d.c. level of a symmetrical or non-symmetrical square wave can be changed by passing it through a simple C–R differentiating network with a selected time constant. In the examples shown the input waveform has a total period of 1 ms (= 1 kHz), and the C–R network gives time constants of 100 ms, 1 ms, or 0.01 ms.

In *Figure 9.1(a)* the C–R time constant is very long relative to the input waveform period, so the output waveform shape is identical to the input but is level shifted so that it swings about the zero-volts value.

Special waveform generators 165

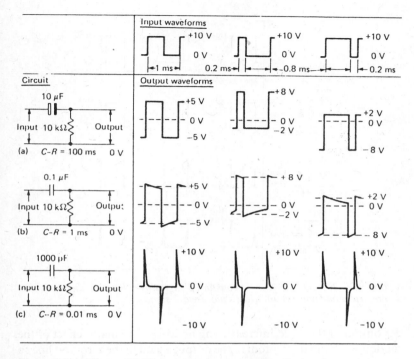

Figure 9.1 *Effects of C–R time constants on symmetrical and non-symmetrical square-wave signals*

Note that the output negative peak/positive peak ratio is proportional to the input signal's mark/space ratio.

In *Figure 9.1(b)* the C–R time constant is equal to that of the input waveform period. The output level again swings about the zero-volts value, but in this case the waveform is modified as well. Note in this circuit that the peak-to-peak amplitude of the output signal is greater than that of the input.

In *Figure 9.1(c)* the C–R time constant is very short relative to the input waveform period. In this case the output signal swings symmetrically about the zero-volts value, irrespective of the input signal's mark/space ratio. The output differs greatly from the input, and takes the form of two spikes or pulses (one positive, the other negative) per input cycle; the peak-to-peak amplitude of the output signal is double that of the input waveform.

Figure 9.2 shows what happens to the output waveforms of *Figure 9.1* when a positive clamping or discriminating diode is wired across

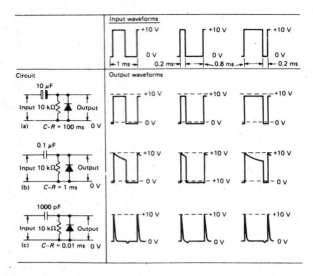

Figure 9.2 *Effects of C–R time constant plus positive clamping diode on symmetrical and non-symmetrical square-wave signals*

the output of the *C–R* differentiating network. The major effect of the diode is to clamp the bottom of the output signal to the zero-volt line so that only positive output signals are obtained. Note in *Figure 9.2(c)* that the negative output spike is virtually eliminated, so that only a single positive output spike is obtained from each input cycle.

Figure 9.3 shows the effect of changing the polarity of the clamping or discriminating diode of the *Figure 9.2* circuit. In this case the tops of the output signals are clamped to the zero-volts line and only negative output signals are obtained.

Figure 9.4 shows how symmetrical sine or square waveforms can be modified by passing them through a long time-constant differentiating network combined with various diode configurations. The *C–R* network has a time constant of 100 ms, compared to the input signal period of 1 ms.

The *Figure 9.4(a)* circuit consists of a *C–R* network only, and simply shifts the output signal level so that it swings symmetrically about the zero-volts value.

The *Figure 9.4(b)* circuit has a positive clamping diode wired across its output. The effect is to clamp the bottom of the output waveform to the zero volts line and give positive outputs only. In the case of the square wave the output is simply level-shifted and its amplitude

Special waveform generators 167

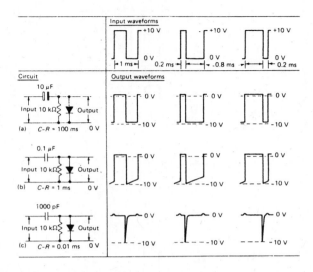

Figure 9.3 *Effects of C–R time constant plus negative clamping diode on symmetrical and non-symmetrical square-wave signals*

roughly equals that of the input, but in the case of the sine wave the bottom half of the waveform is virtually chopped off.

The *Figure 9.4(c)* circuit has a negative clamping diode wired across its output, and this has an effect that is the reverse of that of *Figure 9.4(b)* and gives negative outputs only.

The *Figure 9.4(d)* circuit has a positively biased negative diode across its output. The diode is biased at $+2\,V$ in the example shown, and the effect is to clamp the top of the waveform to the $+2\,V$ reference level. When a square wave is fed through this circuit its mean levels are shifted so that its output switches between $+2\,V$ and $-8\,V$, but when a sine wave is fed through it the lower half is unaltered but its top half is clipped at $+2\,V$. The reverse effect can be obtained by simply changing diode polarity.

Figure 9.4(e) shows the effect of combining the *clamping diode* of the *Figure 9.4(b)* circuit with the *clipping diode* of the *Figure 9.4(d)* circuit. This causes the lower half of the output waveform to be clamped to just below zero volts and the top half to be clipped at $+2\,V$. Note the sine-wave output is greatly distorted.

Figure 9.4(f) shows the effect of combining the positively biased *negative diode* of *Figure 9.4(d)* with a negatively biased *positive diode*. Each diode is biased at $2\,V$, and this causes the output signals to swing

168 Timer/Generator Circuits Manual

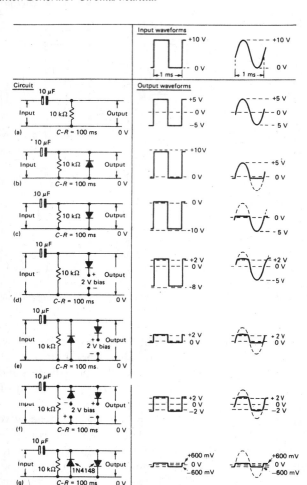

Figure 9.4 *Effects of long time constant C–R network plus clamping and/or clipping diodes on symmetrical sine and square waveforms*

symmetrically about the zero volts reference value but be clipped at $+2\,V$ and $-2\,V$.

Figure 9.4(g) shows the effect of wiring both *positive* and *negative silicon clamping diodes* across the output of the *C–R* network. This circuit is like *Figure 9.4(f)* but has zero bias on each diode. This makes the output swing symmetrically about the zero-volts reference value but clips it at the silicon diode conduction values of about +600 mV and −600 mV.

Finally, *Figure 9.5* shows the effects of placing simple rectifier diodes in series with the outputs of the C–R networks. In each case the C–R network simply causes the diode input signals to swing symmetrically about zero volts, and the diode rectifies or cuts off part of the output signal. The *Figure 9.5(a)* circuit cuts off the whole bottom half of the

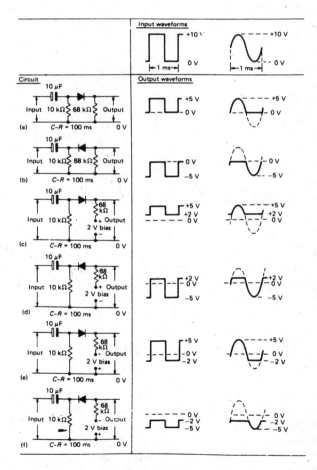

Figure 9.5 *Effects of long time constant C–R network plus series output diode on symmetrical sine and square waveforms*

waveform and gives a positive output, and the *Figure 9.5(b)* one cuts off the top half and gives a negative output.

In the *Figure 9.5(c)* and *(d)* circuits a bias voltage is effectively applied to the series diodes and controls the point at which diode conduction can start. Bias values of +2 V are used, and the action of the *Figure 9.5(c)* circuit is such that all signals below this +2 V reference level are eliminated and only those parts in excess of +2 V appear at the output. The action of the *Figure 9.5(d)* circuit is such that all signals above +2 V are eliminated and only those below +2 V appear at the output. Alternative clipper action can be obtained by applying negative bias to the diodes, as shown in *Figure 9.5(e)* and *(f)*. The clipping levels of these four circuits can be changed by using alternative bias voltage values.

Note that all waveforms shown in the *Figure 9.2* to *9.5* circuits are approximate only, and do not necessarily take precise account of the effects of diode volt drops or the effects of input signal-source impedance. Also note that a variety of useful waveforms can be obtained from all these circuits by feeding them with triangle or sawtooth inputs, or by feeding their outputs to simple filter networks.

White noise generators

White noise can be described as a signal containing a full spectrum of randomly generated frequencies, all with randomly determined amplitudes, but which have equal power *per bandwidth unit* when averaged over a reasonable unit of time. White noise is of value in testing AF and RF amplifiers and in generating special sound-effects, etc.

White noise can be generated using either analogue or digital techniques. Zener diodes act as excellent sources of analogue white noise, and *Figure 9.6* shows a simple but useful circuit that uses one of these as its basic noise source. Here, R_1 and the zener diode form a d.c. negative feedback loop between the collector and base of common-emitter amplifier Q_1 and thus stabilize its d.c. working levels, but the loop is a.c. decoupled via C_1, so that the zener acts as a noise source and is in series with Q_1 base. The zener can be any 5.6 V type, and Q_1 simply amplifies its noise and provides a useful (typically about 1 V peak-to-peak) white noise output at its collector.

The base-emitter junction of an ordinary npn transistor acts as a zener diode when reverse biased and can thus be used as an inexpensive white-noise source; that of the 2N3904 typically zeners at about 5.6 V,

Special waveform generators 171

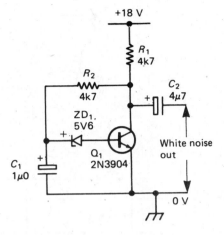

Figure 9.6 *White noise generator*

and *Figure 9.7* shows how it can be used as a zener in the *Figure 9.6* circuit.

Simulated white noise can be generated digitally via a maximum length pseudorandom sequence generator. *Figure 9.8* illustrates the basic principle. Here, an 18-stage shift register is clocked at

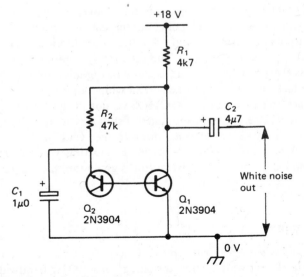

Figure 9.7 *Alternative white noise generator*

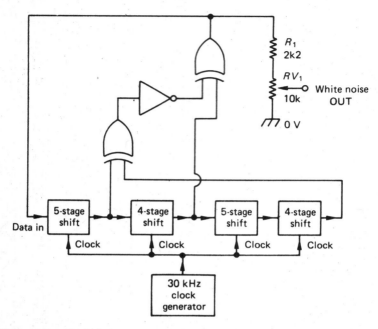

Figure 9.8 *Block diagram of digital white noise generator*

30 kHz and wired so that its data terminal logic is fed forward one step on the arrival of each clock pulse, but this data is derived by EX-ORing the outputs of stages 5, 9 and 18 so that a pseudorandom or jumbled output sequence is generated. This apparently random sequence in fact repeats every few seconds but, because its digital output is very rich in harmonics, otherwise acts like perfect white noise.

Figure 9.9 shows a practical CMOS version of the digital white noise generator. Here, the eighteen stages of shift register are obtained from IC_2, and the 30 kHz clock generator is formed via IC_{1a}–IC_{1b}, which are wired as an astable multivibrator, and EX-ORing is obtained via IC_{1c}–IC_{1d} and Q_1. The white noise output signal amplitude is variable via RV_1.

Pink noise

White noise is a signal with a power (energy) content that is constant at all units of bandwidth. Thus, the energy of a 100 Hz bandwidth white noise signal is the same at 1 kHz as it is at 10 kHz; but 100 Hz band-

Special waveform generators 173

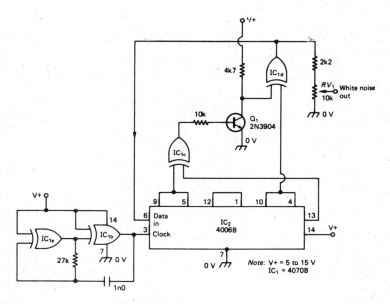

Figure 9.9 *Practical digital white noise generator*

widths are ten times more common at 10 kHz than they are at 1 kHz, and it can thus be seen that (by definition) output voltage levels of a white noise generator rise at a rate of 3 dB per octave. Thus, if you listen to (or measure) white noise signals, they are naturally dominated by high-frequency hiss.

A far more useful noise signal is one in which the output level is flat over the whole frequency band, so that, for example, the frequency responses of an amplifier, filter, or graphic equalizer can be quickly checked be feeding this noise signal to its input and measuring the frequency levels at its output. Noise with this specific characteristic is known as pink noise, and is normally generated by feeding ordinary white noise (as generated by *Figures 9.8* and *9.9*) through an *R–C* filter that is configured to give a first-order slope of −3 dB/octave. *Figure 9.10* shows an add-on filter that is designed for this purpose and gives (when used with any of the above white noise generators) a pink noise characteristic that is within 1/4 dB of this value over the entire 10 Hz to 40 kHz frequency range.

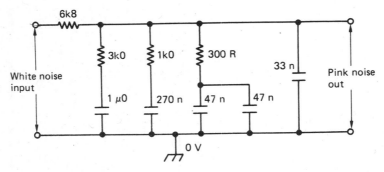

Figure 9.10 *Add-on white-noise to pink-noise converter*

Crystal oscillators

Crystal oscillators generate waveforms with a very high degree of frequency accuracy and stability. They use piezoelectric quartz crystals as high-precision electromechanical resonators or tuned circuits. These crystals have typical Qs of about 100,000 and provide roughly a thousand times greater frequency stability than a conventional *LC* tank circuit. Their operating frequency, which can vary from a few kilohertz to 100 MHz, is determined by the mechanical dimensions of the crystal, which can be cut to provide either series or parallel resonant operation. Series-mode devices present a low impedance at resonance, while parallel-mode devices present a high impedance at resonance.

Figures 9.11 to *9.13* show some practical examples of transistor-based crystal oscillator circuits. The *Figure 9.11* design is based on the Pierce oscillator, and can be directly used with virtually any good 100 kHz to 5 MHz parallel-mode crystal.

Alternatively, *Figure 9.12* shows a 100 kHz oscillator that is designed for use with a series-mode crystal. In this case the circuit is wired as a Colpitts oscillator, and the L_1–C_1–C_2 tank circuit is designed to resonate at the same frequency as the crystal.

Figure 9.13 shows an outstandingly useful two-transistor oscillator circuit that can be used with virtually any 50 kHz to 10 MHz series-resonant crystal. In this design Q_1 is wired as a common base amplifier and Q_2 is an emitter follower, and the output signal (from Q_2 emitter) is fed back to the input (Q_1 emitter) via C_2 and the series-resonant crystal. This really is an excellent circuit, and will oscillate with almost any crystal that shows the slightest sign of life.

Special waveform generators 175

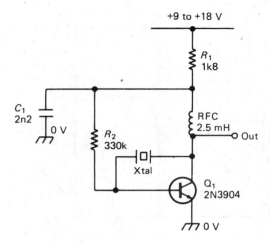

Figure 9.11 *Wide-range Pierce oscillator with parallel-mode crystal*

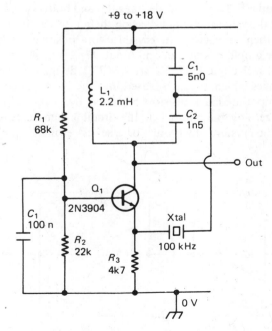

Figure 9.12 *100 kHz Colpitts oscillator using series-mode crystal*

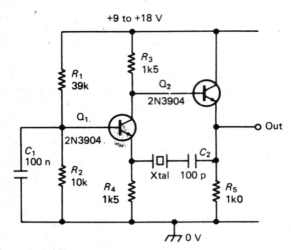

Figure 9.13 *Wide-range (50 kHz–10 MHz) oscillator can be used with any series-mode crystal*

Some simple TTL or CMOS digital gates and buffers can be made to act as crystal oscillators by first biasing them into the linear amplifier mode and then connecting the crystal into a positive feedback path between the amplifier's output and input. *Figures 9.14* and *9.15* show examples of such circuits. The *Figure 9.14* TTL design uses two 74LS04 inverter stages which are each biased into the linear mode via 470 Ω output-to-input feedback resistors and then a.c. coupled in series via C_1, to give zero overall phase shift; the circuit is then made to oscillate by wiring the crystal (which must be a series-resonant type) between

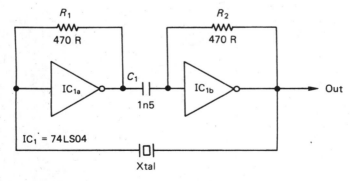

Figure 9.14 *TTL-based crystal oscillator*

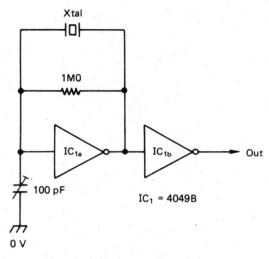

Figure 9.15 *CMOS-based crystal oscillator*

the output and input. This circuit can operate from a few hundred kilohertz to above 10 MHz, but the C_1 value may need adjustment.

Finally, the *Figure 9.15* CMOS circuit is based on a pair of 4049B inverter stages. The first stage is used as a crystal oscillator by wiring it into the linear amplifier mode via R_1 and feeding the output back to the input via the parallel-resonant crystal. The second stage is used as a simple output buffer.

Linear staircase generator

A linear staircase generator circuit has both input and output terminals, and its basic action is such that its output starts at a low level but then rises by a discrete step each time an input pulse is applied, until eventually, after a predetermined number of input cycles, the output switches abruptly back to the low level and the whole sequence starts to repeat. The output thus takes the form of a staircase with a predetermined number of steps. Staircase generators can thus be used as pulse counters, frequency dividers, or step-voltage generators for use in transistor curve tracers.

Figure 9.16 shows a practical linear staircase generator circuit. Here, Q_1 is wired as a simple common emitter amplifier, which controls constant-current generator Q_2, which controls the charging current of

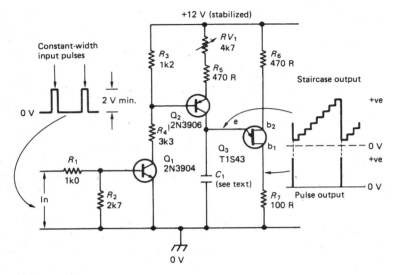

Figure 9.16 *Linear staircase generator circuit*

capacitor C_1, which is coupled to the input of unijunction transistor Q_3. Normally the circuit's input is low, so Q_1 and Q_2 are cut off and no charge is fed to C_1. Each time that a constant-width positive input pulse arrives Q_1 and Q_2 are driven on and a charge current is fed into C_1, which charges linearly for the duration of the pulse. The C_1 voltage thus increases by a fixed amount each time an input pulse is applied, until eventually, after a predetermined number of pulses, the C_1 voltage reaches the trigger value of Q_3 at which point the UJT fires and discharges C_1, thus restarting the operating cycle.

If the input pulses of this circuit are applied at a constant frequency a linear staircase waveform is developed across C_1, and a brief output pulse appears across R_7 each time the UJT fires. If the input frequency is not constant a non-linear staircase is developed across C_1, but a brief output pulse again appears across R_7 after a predetermined number of input pulses have been applied. Stable count/division ratios or staircase steps from two to about twenty can be obtained.

Note that this circuit must be fed with constant-width input pulses that have a width that is small relative to the pulse repetition period. The C_1 value is determined by these considerations, and is best found by trial and error. The division ratio is variable over a 10:1 range via RV_1. Finally, note that the staircase output waveform is available

Special waveform generators 179

across C_1 at a high impedance level; it can be made available at a low impedance level by interposing a Darlington emitter follower buffer stage between C_1 and the final output terminal of the circuit.

Digital sine-wave synthesizer

A sine wave can be created digitally by first building up the rough sine wave outline in a number of digital steps, and then removing the digital

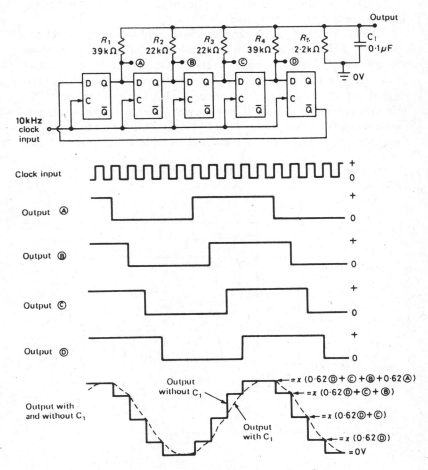

Figure 9.17 *Basic circuit and waveforms of 1 kHz digital sine wave synthesizer*

signal's high frequency components via a simple filter network. *Figure 9.17* illustrates the basic technique.

Here, a clock signal is fed to the input of a five-stage walking ring or Johnson counter. Four of the counter outputs are added together via a resistor weighing network to produce a crude sine wave which is then converted into a reasonably pure form via low-pass filter C_1. Note that the sine-wave output frequency is one-tenth of that of the original clock signal. Consequently, since digital signals generate only odd harmonics, the lowest harmonic of any consequence to the final sine-wave signal are the ninth, eleventh, nineteenth, twenty-first, and so on, and these are easily removed via C_1.

Finally, to complete this chapter, *Figure 9.18* shows the circuit of a practical 1 kHz digital sine wave synthesizer of the type described above. This is built around a 4018B CMOS presettable divide-by-N counter, with transistor Q_1 used to convert an external 10 kHz input signal into a form suitable for clocking the IC. The performance of this circuit has been fully checked on a wave analyser, which shows that the lowest significant harmonic of the 1 kHz signal is the ninth at $-36\,dB$ relative to the fundamental. The sine wave thus has a total harmonic

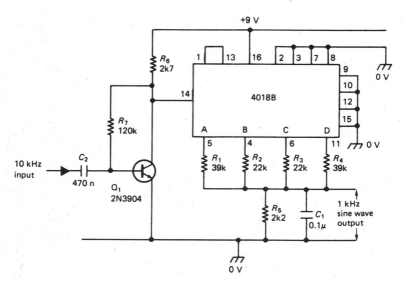

Figure 9.18 *Practical 1 kHz digital sine-wave synthesizer*

distortion content of about 2%. If a second order low-pass filter is used in place of C_1, all harmonics are reduced to better than 65 dB down on the fundamental, giving a THD value of about 0.1%. The *Figure 9.18* circuit thus provides a simple and inexpensive means of generating good quality sine waves.

10 Phase-locked loop circuits

Phase-locked loop (PLL) devices and techniques are widely used in applications such as automatic frequency tracking, frequency multiplication, and frequency synthesis, etc. The basic principles of PLL operation have already been fully described in Chapter 1, where it was also pointed out that the three best known PLL ICs are the **4046B CMOS chip** and the **NE565** and **NE567 devices** from Signetics, which each house excellent VCOs that are outstandingly useful in their own right. In this chapter we take a detailed look at all three of these ICs.

The 4046B IC

The 4046B CMOS IC is an outstandingly useful PLL. Its built-in VCO can be voltage-scanned through a million-to-one frequency range and has a top-end frequency limit in excess of 1 MHz. *Figure 10.1* shows the outline and pin notations of the device, which is housed in a 16-pin DIL package. This figure also shows the internal block diagram of the 4046B and the basic external connections of its integral VCO.

The 4046B houses two different types of phase comparator, plus an excellent VCO, a zener diode, and a source-follower buffer stage. Phase comparator 1 (PC_1) is a simple EX-OR type and has a good noise-rejection performance, but must be driven by square waves on both pins 3 and 14. It has only a narrow capture-frequency range. Phase comparator 2 (PC_2) is an edge-triggered logic/bistable type with a three-state output and can be driven by grossly non-symmetrical waveforms on pins 3 and 14. It has a very wide capture-frequency range, but fairly poor noise rejection.

The 4046B's VCO is a wide-range type with a maximum operating

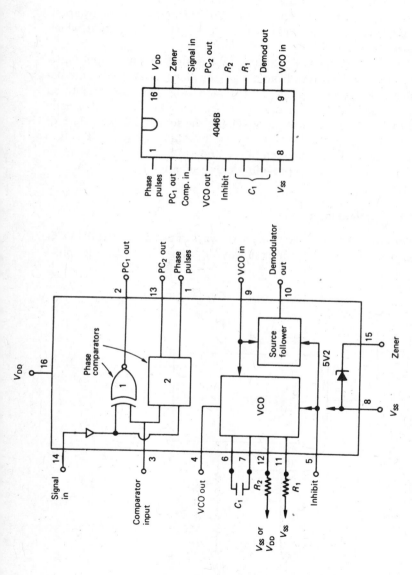

Figure 10.1 *Internal block diagram and pin-outs of the 4046B PLL IC*

frequency in excess of 1 MHz. The frequency is determined by the voltage on pin 9, by the capacitor value between pins 6 and 7 (50 pF minimum), and by the values of R_1 and R_2. R_2 enables the minimum operating frequency to be preset, and can be eliminated in many applications. The VCO generates a symmetrical square-wave output, which appears in pin 4.

The IC's pin 9 VCO-input terminal has a near infinite input impedance and can be driven from a high-impedance source. The internal source-follower enables the pin 9 voltage to be externally monitored without loading this source. Inhibit terminal 5 of the IC is normally tied to V_{ss} to enable both the VCO and the source follower. Both of these devices are disabled when a logic-1 is applied to pin 5. The IC's internal zener diode (between pins 8 and 15) has a nominal value of 5V6, and can be used to provide supply regulation if required.

VCO applications

Figures 10.2 to *10.10* show some basic ways of using the 4046B's VCO. *Figure 10.2* shows the simplest way of using the VCO. The pin 9

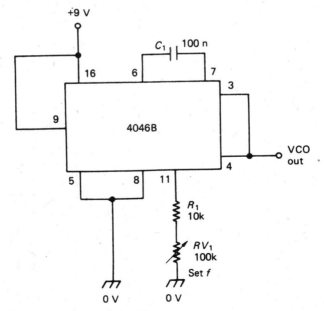

Figure 10.2 *Simple 200 Hz–2 kHz square-wave generator*

voltage-control input is tied permanently high and the circuit acts as a basic square-wave oscillator with its frequency variable over a 10:1 range via RV_1. Note at this point that the VCO output (pin 4) is tied directly to the pin 3 phase comparator input: if pin 3 is allowed to float the comparators self-oscillate at about 20 MHz and superimpose an HF signal on the top part of the VCO output waveform.

Figure 10.3 shows the 4046B wired as a wide-range VCO. R_1–C_1 determine the maximum frequency that can be obtained and RV_1 controls the actual frequency via the pin 9 voltage. The frequency falls to near-zero (a few cycles per minute) with pin 9 at zero volts. The effective control range of pin 9 varies from roughly one volt above 0 to 1 V below the positive supply value, e.g., RV_1 has a dead control area of several hundred millivolts at either end of its span.

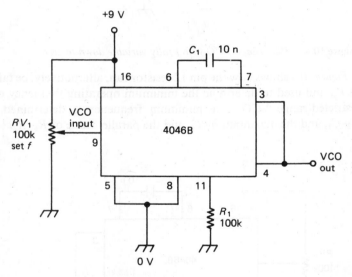

Figure 10.3 *Wide-range VCO, variable from near-zero to 1.4 kHz via the pin 9 voltage*

Figure 10.4 shows how these dead areas of RV_1 can be eliminated by wiring a silicon diode in series with each end of RV_1, and how the minimum operating frequency can be reduced to zero by wiring high-value resistor R_2 from pin 12 to V_{DD}. Note that when the frequency is reduced to zero the VCO output randomly settles in either the logic-0 or logic-1 state.

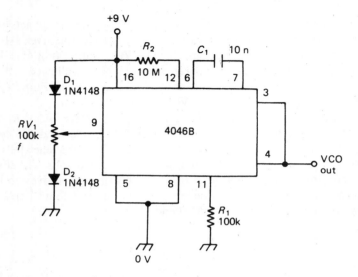

Figure 10.4 *Wide-range VCO with f fully variable down to zero*

Figure 10.5 shows how the pin 12 resistor can, alternatively, be taken to V_{ss} and used to determine the minimum operating frequency of a restricted-range VCO. The minimum frequency is determined by R_2–C_1, and the maximum by C_1 and the parallel value of R_1–R_2.

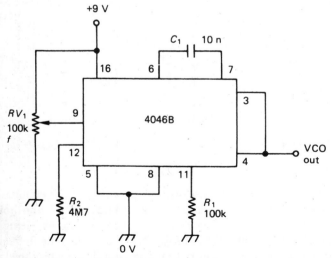

Figure 10.5 *Restricted-range VCO, variable from 60 Hz–1.4 kHz via RV_1*

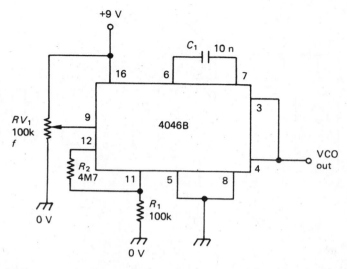

Figure 10.6 *Alternative version of the restricted range VCO*

Figure 10.6 shows an alternative version of the restricted-range VCO, in which the maximum frequency is controlled by R_1–C_1, and the minimum by C_1 and $R_1 + R_2$. Note that, by suitable choice of R_1 and R_2 values, the restricted-range VCO can be made to span any range from 1:1 to near-infinity.

The VCO can be made to generate a pair of anti-phase square-wave outputs by connecting the VCO output to the phase-comparator input, taking the signal input (pin 14) high, and taking the antiphase output from pin 2, as shown in *Figure 10.7*. Note that this circuit makes use of the IC's built-in EX-OR gate (PC_1).

The VCO section of the 4046B can be disabled by taking inhibit pin 5 high (to logic-1). This feature enables the VCO to be gated on and off via external signals. *Figure 10.8* shows how the VCO can be manually gated via a push button switch wired to pin 5, and *Figure 10.9* shows how the circuit can be gated electronically via an external inverter stage (made from a 4011B CMOS gate). Alternatively, if the two-phase output facility is not needed, *Figure 10.10* shows how the internal EX-OR phase detector (PC_1) can be used to give gate control. Note in this latter case that pin 4 is not connected to pin 3.

That completes our look at the basic features of the 4046B's VCO section. Let us now move on and look at how some of these features can be combined or used to make really useful practical circuits.

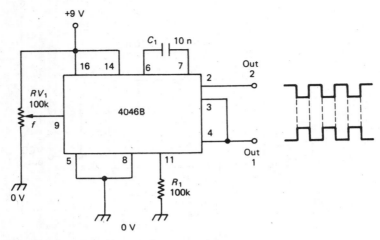

Figure 10.7 *A two-phase wide-range VCO*

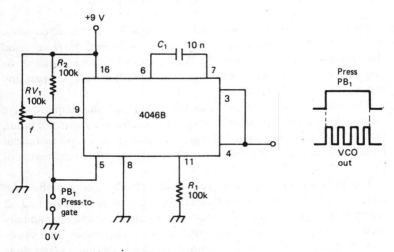

Figure 10.8 *Manually-gated wide-range VCO*

Sound-effects generators

The VCO section of the 4046B IC is exceptionally versatile. Its operating frequency can be scanned over a very wide range via a control voltage applied to pin 9, and its output can be gated on or off via a

Phase-locked loop circuits 189

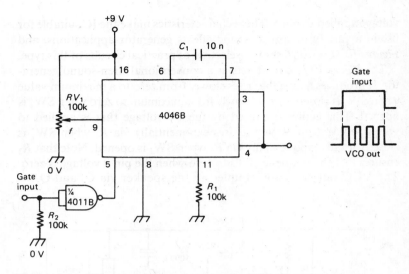

Figure 10.9 *Electronically-gated wide-range VCO using an external gate inverter*

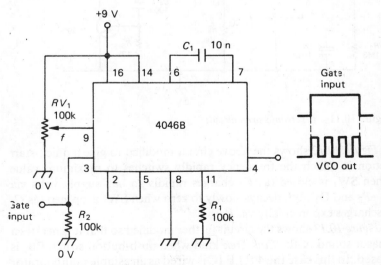

Figure 10.10 *Electronically gated wide-range VCO using the internal EX-OR phase detector for gate inversion*

voltage applied to pin 5. These characteristics make the IC suitable for use in a variety of special sound-effects generator applications, and *Figures 10.11* to *10.16* show a selection of practical circuits of this type.

The *Figure 10.11* circuit acts as a conventional siren-sound generator. It produces a tone that rises slowly from zero to a maximum value when SW_1 is closed, or falls slowly from maximum to zero when SW_1 is opened. This action is caused by the C_1 voltage that is applied to voltage-control pin 9, which rises exponentially via R_1 when SW_1 is closed, or falls exponentially via R_2 when SW_1 is opened. Note that R_3 ensures that the frequency falls to zero when the pin 9 voltage is zero. The VCO output is a.c. coupled to the speaker via C_4 and Q_1.

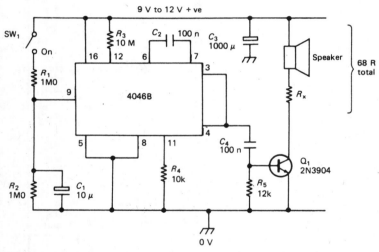

Figure 10.11 *Electronic siren circuit*

Figure 10.12 shows the above circuit modified to give a quick-start action in which the frequency rapidly switches to maximum value when SW_1 is closed (as C_1 charges rapidly to half-supply volts via R_1–R_2 and D_1), but decays slowly to zero when SW_1 is opened (as C_1 discharges exponentially via R_3).

Figure 10.13 shows the circuit further modified so that it generates a phasor sound of the *Star Trek* kind when push-button switch PB_1 is closed. In this case the 4011 B IC is wired as an astable multivibrator that is gated via PB_1 and produces a chain of 4 ms pulses at intervals of 70 ms. Each pulse rapidly charges C_2 via R_3–D_2, to produce a high tone

Phase-locked loop circuits 191

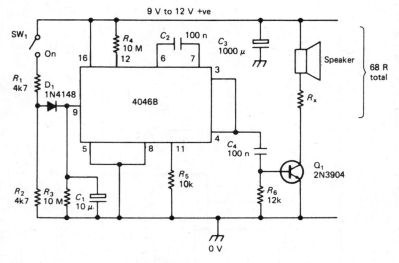

Figure 10.12 *Quick-start siren circuit*

that then decays fairly slowly as C_2 discharges via R_5, only to be repeated again on the arrival of the next pulse.

A different type of sound generator circuit is shown in *Figure 10.14*. This design can be used to generate either a pulsed tone or a warble tone signal (depending on the SW_1 setting) when PB_1 is closed. PB_1 is used simultaneously to enable pin 5 of the 4046B and to gate on the 4001 B astable multivibrator, which then applies a rectangular (alternately fully-high and fully-low) waveform to pin 9. In the pulsed mode, the VCO generates zero frequency when pin 9 is low. In the warble mode it generates a tone that is 20% down on the high tone when pin 9 is low.

Figure 10.15 shows a special-effects run-down clock/sound generator of the type used in electronic dice and roulette games. The circuit action is such that a fast-spinning clicking sound is generated when PB_1 is pressed, and the clicking rate slowly decreases (runs down) to zero when PB_1 is released. The circuit operates as follows:

When PB_1 is pressed, C_1 rapidly charges to a high voltage via D_2; simultaneously, Q_1 is biased on via D_3–R_4 and connects R_6 between pin 11 and ground, thus making the VCO operate at tens of kilohertz and effectively generate an unpredictable (random) number of clock pulses. When PB_1 is released, Q_1 turns off and VCO timing is governed by R_7; simultaneously, C_1 rapidly discharges to half-supply volts via

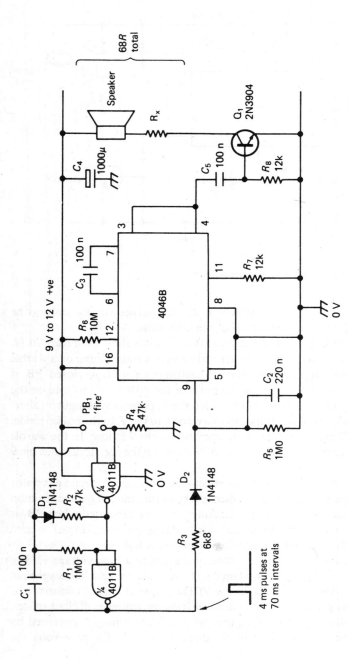

Figure 10.13 *Phasor-sound generator circuit*

Phase-locked loop circuits 193

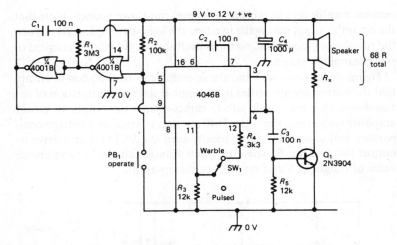

Figure 10.14 *Combined pulsed-tone/warble-tone alarm generator*

R_1–R_2–D_1, causing the VCO to operate at about 100 Hz. C_1 then slowly discharges via R_3, and the VCO frequency slowly decays to zero over a period of about 15 s.

The output of the *Figure 10.15* circuit can be used directly to clock most types of counter, and can be direct-coupled (via R_9) to crystal or

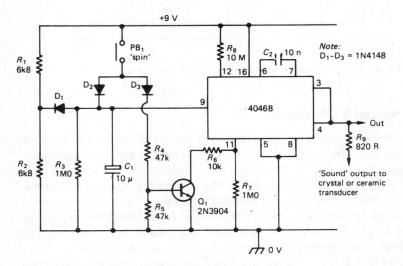

Figure 10.15 *Run-down clock/sound generator circuit*

ceramic transducers to produce low-level run-down sounds. Note that the circuit's output may settle in either the logic-0 or logic-1 state when the run-down is complete, so the output should not be d.c.-coupled to power amplifier stages.

Figure 10.16 shows how the above circuit can be modified to ensure that the output always settles in the logic-0 state on completion of the run-down, thus making it safe to direct-couple the output to power amplifier stages, etc. Here, the 3140 op-amp is wired as a voltage comparator and is used to automatically turn the VCO off and drive its output low (via pin 5) when the pin 9 voltage falls below a reference value of roughly 2 V (set on pin 3 of the op-amp).

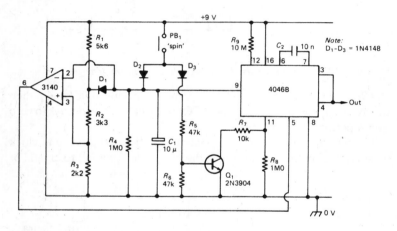

Figure 10.16 *Modified version of the run-down generator*

Special VCO circuits

The versatility of the VCO section of the 4046B makes it suitable for use in a variety of special-purpose waveform generator applications, and *Figures 10.17* to *10.19* show a brief selection of such circuits.

The *Figure 10.17* circuit is that of a simple FSK square-wave generator. With the particular component values shown, this circuit generates a tone frequency of 2.4 kHz when a logic-1 signal is applied to pin 9, or a 1.2 kHz tone when a logic-0 signal is applied to the same point. The high tone is determined by the C_1–R_2 values, and the low tone by the C_1 and $R_2 + R_3$ ones. Alternative tone frequencies can readily be obtained by altering these component values.

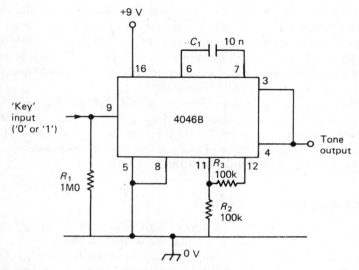

Figure 10.17 *FSK generator. Logic-0 = 1.2 kHz; logic-1 = 2.4 kHz*

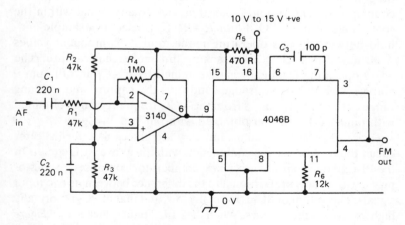

Figure 10.18 *220 kHz FM generator*

The *Figure 10.18* circuit is that of a 220 kHz FM waveform generator. Here, the internal zener (pin 15) of the 4046B is used to provide a stable supply to the 3140 op-amp, which is wired as a ×20 inverting a.c. amplifier but has a quiescent bias of about 2.6V applied to its non-inverting (pin 3) input via $R_2 - R_3$, so that the output (pin 6) of the op-amp comprises a mean 2.6 V potential that is amplitude modulated with an amplified ($\times 20$) version of the AF input signal. This output is applied to the voltage-control input terminal (pin 9) of the 4046B's VCO, which has its C_3–R_6 component values chosen so that it generates a mean output carrier frequency of 220 kHz, which is frequency modulated via the original AF input signal.

Finally, *Figure 10.19* shows how the 4046B VCO can be used as a wide-range universal square-wave clock generator that spans the nominal range 0.5 Hz to 500 kHz in three switch-selected bands. This simple but very useful piece of test gear provides a two-phase output and can be used in either the free-running or the gated modes.

Phased-locked loop circuits

The 4046B can be used in a variety of PLL applications. *Figure 10.20* shows it used as a wide-range signal tracker that will capture and track any input signal within the approximate frequency range 100 Hz to 100 kHz, provided that the pin 14 input signal switches fully between the 0 and 1 logic levels. Note that this circuit (and all others shown in this section) makes use of wide-range phase-comparator 2 (PC_2), and it can thus lock to any signal within the span range of the VCO. Filter R_2–R_3–C_2 is used as a sample-and-hold network in this operating mode, and its component values determine the settling and tracking times of signal capture. The VCO operating frequency is determined by R_1–C_1 and the pin 9 voltage; the VCO span range (and thus the capture and tracking range of the circuit) ranges from the VCO frequency value obtained with pin 9 at 0 V to that obtained with pin 9 at V_{DD} volts.

Figure 10.21 shows a simple but very useful **lock detector/indicator circuit** that can be used in conjunction with the above PLL circuit. In the PLL, the output of each phase comparator comprises a series of pulses with widths proportional to the difference between its two input signals. The output of PC_1 is normally low, and that of PC_2 is normally high, except for these pulses. When the PLL circuit is locked (see *Figure 10.20*) the two outputs are almost perfect mirror images of each other; when the loop is not locked the signals are greatly different.

In the *Figure 10.21* lock detector/indicator circuit the above facts are put to use via 2-input NOR gate IC_{1a}, which is driven from the outputs of the two comparators. The circuit action is such that if the loop is locked the IC_{1a} output remains permanently low, thus driving IC_{1b} output high and illuminating LED_1. If the loop is not locked, however,

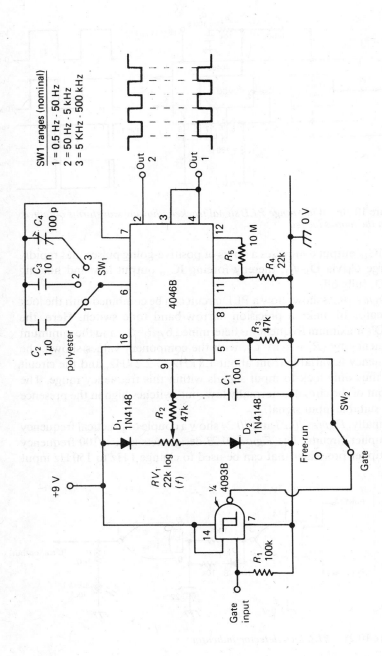

Figure 10.19 Universal clock/square-wave generator

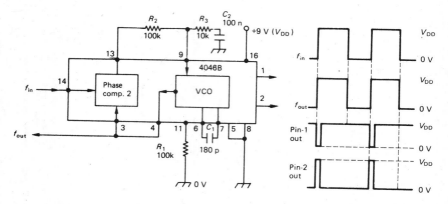

Figure 10.20 *Wide-range PLL signal tracker, showing waveforms obtained when the loop is locked*

the IC_{1a} output comprises a series of positive-going pulses that rapidly charge C_1 via D_1–R_1, thereby forcing IC_{1b} output low and holding LED_1 fully off.

Figure 10.22 shows how a PLL circuit can be combined with the lock indicator to make a precision narrow-band tone switch. Here, the VCO's maximum frequency is determined by R_1–C_1, and the minimum frequency by $(R_1 + R_2) - C_1$; with the component values shown, the frequency is variable from about 1.8 kHz to 2.2 kHz, and the circuit can thus only lock to input signals within this frequency range. The output of the circuit is normally low, but switches high in the presence of a suitable input signal.

Finally, *Figures 10.23* and *10.24* show a couple of practical frequency multiplier circuits. The *Figure 10.23* design acts as a × 100 frequency multiplier/pre-scaler that can be used to change 1 Hz to 150 Hz input

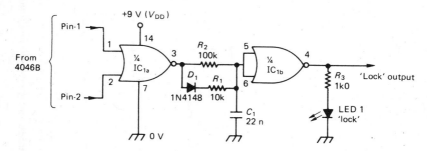

Figure 10.21 *PLL lock detector/indicator*

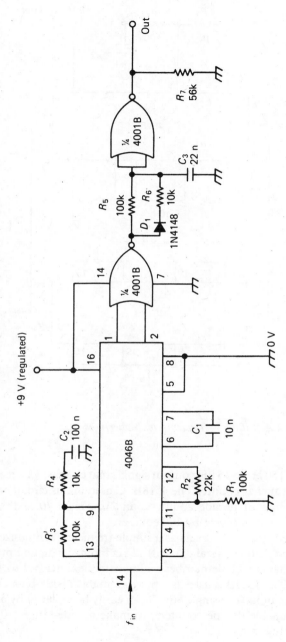

Figure 10.22 *Precision narrow-band (1.8 kHz–2.2 kHz) tone switch*

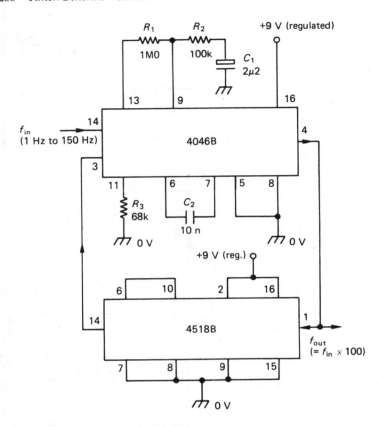

Figure 10.23 *A ×100 low-frequency multiplier/prescaler*

signals into 150 Hz to 15 kHz output signals that can easily be read on a standard frequency counter. The 4518B IC used in this circuit actually contains a pair of decade counters, and in *Figure 10.23* these are cascaded to make a divide-by-100 counter.

The *Figure 10.24* circuit acts as a simple frequency synthesizer. It is fed with a precision (crystal derived) 1 kHz input signal, and provides an output that is a whole-number multiple (in the range ×1 to ×9) of this signal. The 4017B is used as a programmable divide-by-*n* counter in this application. The single 4017B can easily be replaced by a string of programmable decade counters, to make a wide-range (10 Hz to 1 MHz) synthesizer.

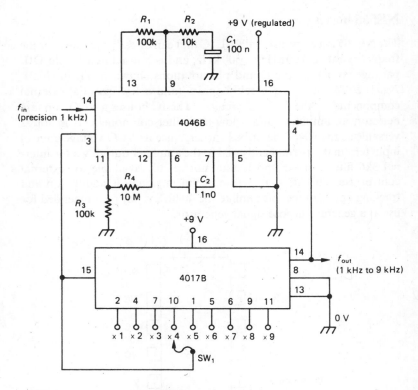

Figure 10.24 *Simple 1 kHz–9 kHz frequency synthesizer*

NE565/567 ICs

The Signetics Corporation manufacture a family of phase-locked loop ICs and associated devices. The three best-known members of this family are the NE565 general-purpose PLL IC, a fairly simply device that can be used in a variety of signal demodulating and locking applications, the NE566 function generator IC, which can be used in a variety of waveform generator applications, and the NE567 tone decoder IC, a specialized PLL device intended purely for use in tone decoding/switching applications. The rest of this chapter is devoted to this range of devices.

NE565 basics

The NE565 is a conventional PLL IC that can directly operate over the frequency range 0.001 Hz to 500 kHz, and is housed in a 14-pin DIL package with the outline and pin notations shown in *Figure 10.25*. *Figure 10.26* shows the block diagram (and a few essential external components) of the device's circuitry. The IC houses a VCO, a phase detector, an amplifier, and a low-pass filter component. It is not as versatile as the 4046B described earlier, since its VCO voltage control input terminal is permanently tied to the amplifier output via the internal 3k6 filter resistor and is thus not readily available for external control use. This IC is therefore useful in signal demodulation and tracking applications, but, unlike the 4046B, is not recommended for use as a general-purpose signal generator.

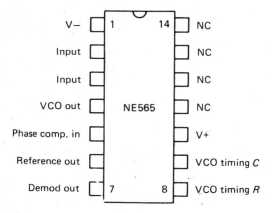

Figure 10.25   *Outline and pin notations of the NE565 PLL IC*

Figure 10.27 shows the basic signal tracking or FM demodulation application diagram of the NE565 when used with a split 12 V (+6 V and −6 V) power supply. In normal use, the external signal that is to be tracked or demodulated is fed to input pin 2 of the phase detector, and unused input pin 3 is signal grounded. The pin 4 output terminal of the VCO is connected to the phase detector's pin 5 input terminal, thus completing the phase locked loop: the VCO's free-running frequency (f_o) is adjusted via the R–C network connected to pins 8 and 9 so that it corresponds to the mid value of the external input signal.

Phase-locked loop circuits 203

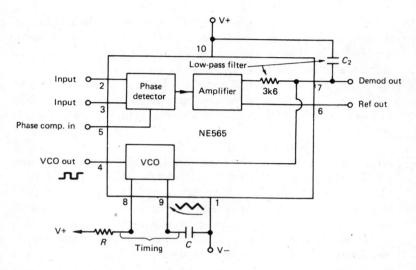

Figure 10.26 *Functional block diagram of the NE565*

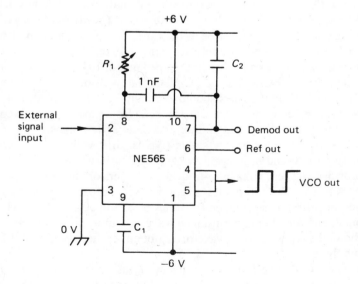

Figure 10.27 *Basic signal tracker/FM-demodulator application diagram*

Under the above conditions the VCO frequency can lock to that of the input signal because (when the lock condition is feasible) the mean d.c. level of the phase detector's amplified output is proportional to the difference between the input and VCO frequencies and is used to voltage control the VCO input. Thus, if the input frequency rises above that of the VCO, the detector's output also rises and automatically forces the VCO frequency to rise towards that of the input until locking occurs.

Note that in practice a small time delay is applied to this locking action via the single-pole loop filter formed by C_2 (connected between pins 7 and 10) and the IC's internal 3k6 resistor. Thus, if the input signal is noisy or juddery, or is frequency modulated (FM), the VCO locks to the mean frequency of the input signal and generates a clean output at pins 4 or 5, and produces a demodulated FM output at pin 7.

Note when using the NE565 in practical circuits that a small capacitor (about 1 nF) should be wired between pins 7 and 8, to enhance circuit stability.

More details

Figure 10.28 lists the main parameter and characteristic details of the NE565. The IC is normally used with a split (positive and negative) power supply, which must be in the range 5 to 12 V, but can also be used with a single ended supply in the range 10 to 24 V. The IC's phase detector section has a typical input impedance of 10 k on each terminal, and the IC can lock and track to input signals with amplitudes as low as 1 mV r.m.s. Input signals should normally be a.c. coupled, but can be d.c. coupled if the d.c. resistances seen from pins 2 and 3 are equal and there is no d.c. voltage difference between the pins.

The IC's VCO is a very stable (typical drift with temperature is 300 p.p.m./°C and with supply voltage is 0.2%/V) wide-range type that provides excellent voltage-to-frequency conversion linearity (typically 0.5%). The VCO provides a good TTL-compatible square-wave output (with typical rise and fall times of 20 and 50 ns respectively) at pin 4, and a highly linear triangle wave output at pin 9. *Figure 10.29* shows the typical output waveforms obtained when using a split 12 V supply.

The VCO's free running frequency (f_o) is set by resistor R wired between pin 8 and pin 10 ($=V+$), and by capacitor C wired between pin 9 and pin 1 ($=V-$), and equals (in kilohertz) $(1.2)/4RC$ when R is

Parameter	Values, with ±6 V supply		
	Min.	Typ.	Max.
Supply voltage	±5 V		±12 V
Input impedance Input sensitivity (r.m.s.)	5k0 10 mV	10k 1 mV	
VCO Maximum frequency Drift with temperature Drift with supply volts		500 kHz 300 p.p.m./°C 0.2%/V	1.5%/V
Triangle wave Output amplitude (p-p) Output linearity		2.4 V 0.5%	3 V
Square wave Logic-1 output volts Logic-0 output volts Rise time Fall time Output sink current Output source current	+4.9 V 0.6 mA 5 mA	+5.2 V −0.2 V 20 ns 50 ns 1 mA 10 mA	+0.2 V
Demodulated output Output voltage level (pin 7) Max. output voltage swing Output voltage swing at 10% FM Total harmonic distortion Output impedance Offset volts (pin 6 to pin 7) AM rejection	4.0 V 200 mV p-p	4.5 V 2 V p-p 300 mV p-p 0.4% 3.6k 50 mV 40 dB	5.0 V 1.5% 200 mV

Figure 10.28 *Main parameters of the NE565*

in kilohms and C is in microfarads. R can have any value in the range 2k0 to 20 k (optimum value is about 4k0), and C can have any value at all. In normal use the NE565 will phase-lock to any input signal frequency within plus or minus 60% of the f_o value; this is known as the circuit's *lock range*.

The output section of the IC gives a demodulated output at pin 7, and pin 6 provides a d.c. reference voltage that is close to the d.c. potential of pin 7. If a resistance is wired between pins 6 and 7 the gain of the IC's output stage can be reduced with little change in the d.c.

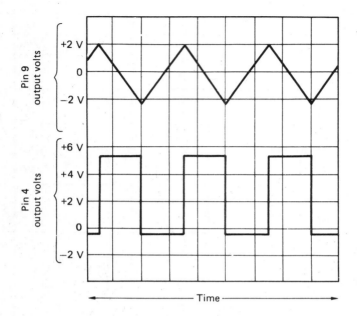

Figure 10.29 *VCO output waveforms using a plus and minus 6 V supply*

voltage level of the output. This allows the lock range to be decreased to as low as 20% of f_o, with little change in the f_o value.

Frequency shift keyed demodulator

Frequency shift keyed (FSK) signals are widely used in binary-data communication systems. At the transmitter the binary signals are used to generate a continuous two-tone carrier signal, with a mark or logic-1 state represented by one tone, and the space or logic-0 state represented by another tone. In the receiver the two-tone carrier is converted back to a binary signal via a precision tone switch or FSK decoder.

Figure 10.30 shows how the NE565 can be used as an FSK decoder of a 1070 Hz/1270 Hz input waveform. As the signal appears at the input, the loop locks to it and tracks it between the two frequencies, with a corresponding d.c. shift at the output. Loop filter C_2 has a small value to eliminate overshoot on the output pulse, and a three-stage *RC* ladder filter is used to remove carrier components from the output. This filter has a band edge roughly half way between the maximum

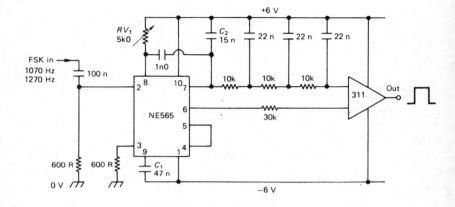

Figure 10.30 *FSK demodulator circuit*

FSK keying rate (300 baud or 150 Hz) and twice the input frequency (about 2200 Hz). The filter output signal is made logic compatible by connecting a 311 voltage comparator between the output and pin 6 of the loop. The free-running frequency of the VCO is adjusted with RV_1 to give a slightly positive output voltage when a 1070 Hz input signal is applied.

Note that the input connection of this circuit is typical for cases where a d.c. voltage is present at the source and a direct connection is thus not allowable. Both input terminals are returned to ground with identical resistors (in this case the values are chosen to give a 600 ohm input impedance).

Single-ended supplies

To conclude this look at the NE565, *Figure 10.31* shows how the IC can be connected as a 60 kHz FM demodulator that is powered from a single-ended 10 to 24 V supply. Here, a resistive voltage divider ($R_1 - R_2$) and R_3 and R_4 are used to apply a balanced bias voltage to the two input terminals (pins 2 and 3), and the 60 kHZ FM signal is a.c.-coupled to input pin 2. The VCO's free running frequency is set to 60 kHz via $R_5 - C_2$ and RV_1, and the decoded output signals are fed through a three-stage low-pass filter to minimize the effects of unwanted noise signals.

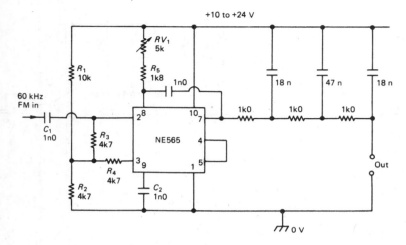

Figure 10.31 *60 kHz FM demodulator using single-ended power supply*

NE566 basics

The NE566 is a general-purpose, wide-range, voltage-controlled function generator IC that simultaneously generates very stable high-quality square and triangle output waveforms, at fixed or variable frequencies up to a maximum of about 1 MHz. These waveforms can easily be frequency modulated (FM) or frequency shift keyed (FSK) by an external signal via a voltage-control input terminal.

The NE566 is an easy device to use. It is housed in an 8-pin DIL package with the outline and pin notations shown in *Figure 10.32*. *Figure 10.33* shows the device's block diagram, plus a few essential

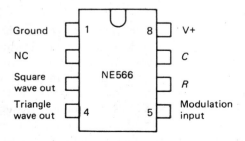

Figure 10.32 *Outline and pin notations of the NE566 function generator IC*

external components, and *Figure 10.34* shows a simple fixed frequency application circuit for the device, which can be powered by any single-ended or split supply voltage in the range 10 to 24 V.

In essence, the NE566 is simply a VCO with buffered outputs. As can be seen from *Figure 10.33*, the VCO section is actually made up of a pair of voltage-controlled current sources that are used to linearly charge or discharge an external timing capacitor, and a Schmitt trigger, which is used to flip the current sources when the capacitor voltage reaches preset levels. Thus, a linear triangle wave is generated across the capacitor, and a high-quality square wave is generated at the Schmitt trigger output. These waveforms are fed to the outside world via simple buffer amplifiers.

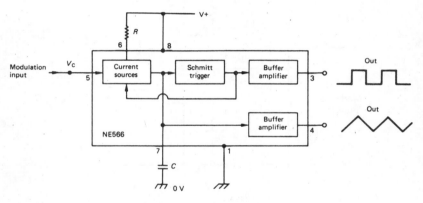

Figure 10.33 *Functional block diagram of the NE566*

More NE566 details

The NE566's operating frequency is determined by an external resistor and capacitor and by the voltage applied to its control terminal. The resistor must have a value in the range 2k0 to 20 k; the capacitor can have any value, and the control voltage must be in the range 75% to 100% of the IC's supply voltage value. The frequency can thus be varied over a 10:1 range via the resistor. It can be varied or modulated over a similar range via the control voltage. Thus, in the simple application circuit of *Figure 10.34* the IC is wired as a fixed frequency FM waveform generator. Note that a 1 nF capacitor is wired between pins 5 and 6, to enhance circuit stability.

The operating frequency of the NE566 is roughly equal to $2(V+ - V_c)/R.C.V+$. Put in a simpler way, *Figure 10.35* shows the span of operating frequencies given at different C values by the *Figure 10.34* circuit when R has a value of 4k0; the frequency varies from 5 Hz at 10 μF to about 200 kHz at 100 pF. *Figures 10.36* and *10.37* show how, when the C value is fixed, the frequency varies when the control voltage or the R value is varied.

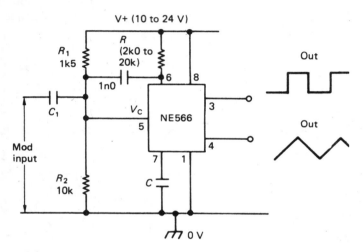

Figure 10.34 *Simple fixed frequency application circuit*

Figure 10.38 lists the main parameter values of the NE566. Note that both output waveforms are available at a very low impedance level (about 50 Ω). *Figure 10.39* shows the typical shapes and voltage levels of the output waveforms when using a 12 V supply.

Finally, to complete this look at the NE566, *Figure 10.40* shows some modifications that can be made to the basic *Figure 10.34* design to convert it into a wide-range three-band FM generator. The frequency is fully variable via RV_1 and switch variable via range selector SW_1, and the triangle and square-wave output levels are fully variable via RV_2 and RV_3. The waveforms can be frequency modulated by applying the modulation waveform to pin 5 via C_1. Note that R_3 is used to raise the circuit's input impedance to about 22 k.

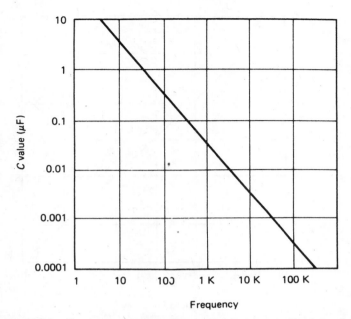

Figure 10.35 *Frequencies given by Figure 10.34 at various C values when $R = 4k0$*

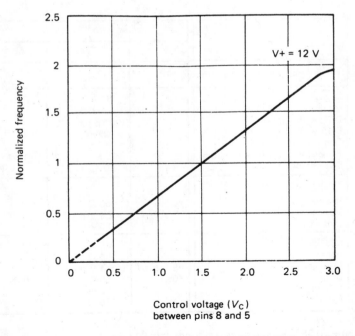

Figure 10.36 *Normalized frequency of the Figure 10.34 circuit as a function*

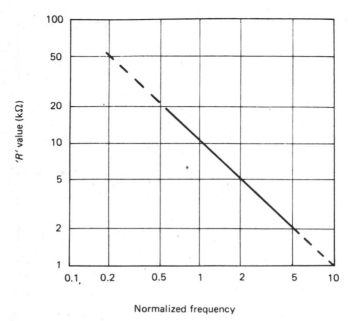

Figure 10.37 *Normalized frequency of the Figure 10.34 circuit as a function of R value*

Parameter	Values, with 12 V supply		
	Min.	Typ.	Max.
Supply voltage	10 V		24 V
VCO Maximum frequency Drift with temperature Drift with supply volts Control terminal input Z FM distortion (±10% deviation) Sweep range		1 MHz 200 p.p.m./°C 2%/V 1 MΩ 0.2% 10:1	1.5%
Triangle wave output Impedance Voltage Linearity	 2 V p-p	50 Ω 2.4 V p-p 0.5%	
Square wave output Impedance Voltage Duty cycle Rise time Fall time	 5 V p-p 40%	50 Ω 5.4 V p-p 50% 20 ns 50 ns	 60%

Figure 10.38 *Main parameters of the NE566*

Phase-locked loop circuits 213

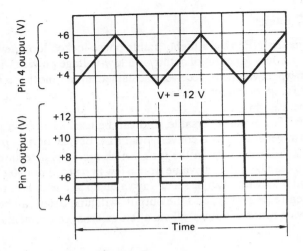

Figure 10.39 *VCO output waveforms*

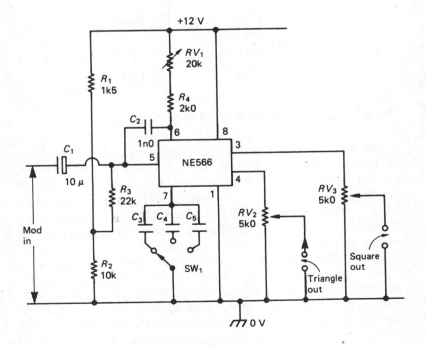

Figure 10.40 *Wide-range three-band FM generator*

NE567 basics

The Signetics NE567 is a highly stable phase-locked loop device that has the primary function of acting as a low-voltage power switch that turns on whenever the IC receives a sustained input tone that is within a narrow range of preset frequency values, i.e., it acts as a precision tone-operated switch.

The NE567 is fairly versatile, and can be used as either a variable waveform generator, as a conventional PLL device, or as a precision tone switch. When used in the tone-switch mode its detection centre frequency can be set at any value from a few hertz to about 500 kHz, its detection bandwidth can be set at any value up to a maximum of 14% of the centre frequency, and the output switching delay can be varied over a wide range of values, all via a few external components.

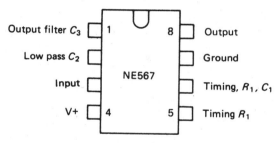

Figure 10.41 *Outline and pin notations of the NE567*

The NE567 is housed in an 8-pin DIL package. *Figure 10.41* shows the outline and pin notations of the IC, and *Figure 10.42* shows the internal block diagram (plus a few essential external components) of the device, which houses a conventional PLL circuit (comprising a VCO, a phase detector, and a feedback filter) plus a quadrature phase detector and an output-driving amplifier and open-collector output transistor. The IC functions as follows:

The VCO section of the IC can be varied over a wide range via external components R_1 and C_1, but can be voltage-controlled over only a very narrow range (a maximum of about 14% of the free-running value) via control pin 2. Consequently, the PLL circuit can lock to only a very narrow range of preset input frequency values. The IC's quadrature phase detector compares the relative frequencies and phases of the input signal and the VCO output, and produces a valid

Phase-locked loop circuits   215

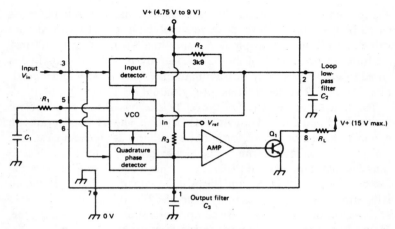

Figure 10.42 *Block diagram and basic external connections for the NE567*

output-driving signal (which turns Q_1 on) only when these two signals coincide, i.e., when the PLL is locked. The centre frequency of the NE567 tone switch is thus equal to the free-running frequency of the VCO, and its bandwidth is equal to the lock range of the PLL.

Figure 10.43 shows the basic connections of the NE567 tone-switch circuit. The input tone signal is a.c. coupled (via C_4) to pin 3, which has

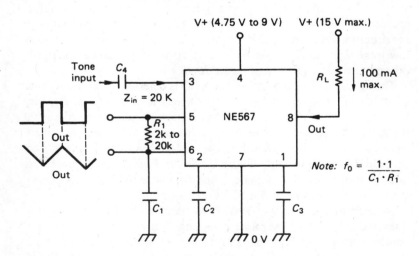

Figure 10.43 *Basic tone switch connections of the NE567*

an input impedance of about 20 k, and the external output load (R_L) is wired between pin 8 and a positive supply voltage with a maximum value of 15 V. Note that pin 8 can sink load currents up to 100 mA maximum value. Pin 7 of the IC is normally grounded, and pin 4 is wired to a positive supply with a minimum value of 4.75 V and a maximum value of 9 V. Ideally, this supply should be stabilized.

The centre frequency (f_o) of the VCO is set via R_1 (any value from 2k0 to 20 k) and C_1, and equals $(1.1)/(R_1.C_1)$. The VCO generates an exponential sawtooth waveform that is available at pin 6, and a square wave that is available at pin 5. The bandwidth of the tone switch (and thus the lock range of the PLL) is determined by C_2 and the internal 3k9 resistor (R_2) of the IC, and the output switching delay of the circuit is determined by C_3 and internal resistor R_3. We will take detailed looks at the selection of each of these components shortly. In the meantime, *Figure 10.44* lists the main parameter values of the NE576 IC.

Voltage controlled oscillator design

The voltage controlled oscillator (VCO) section of the NE567 is quite easy to use, as shown in *Figures 10.45* and *10.46*. It generates a not very useful non-linear ramp waveform on pin 6, and an excellent square wave, with 20 ns rise and fall times, directly on pin 5 of the IC. This square wave has a peak-to-peak amplitude equal to the supply voltage value -1.4 V, and can be externally loaded by any resistance value greater than 1k0 without adverse effects (*Figure 10.45*). Alternatively the square-wave output can be applied (in slightly degraded form) to a low impedance load (at peak currents up to 100 mA) via the pin 8 output terminal, as shown in *Figure 10.46*.

The operating frequency, f, of the VCO is determined by the R_1 and C_1 values, and equals $(1.1)/(R_1.C_1)$, where f is in kilohertz, R_1 is in kilohms, and C_1 is in microfarads. In practice, R_1 must be restricted to the 2k0 tp 20k range. To find the C_1 value needed to generate a given frequency from a given R_1 value, use the formula $C_1 = (1.1)/(f.R_1)$; to help aid the reader in using the VCO section, *Figure 10.47* is presented as a component-selection guide. Thus, to give a 10 kHz VCO operation, C_1/R_1 values of about 55 nF/2k0 or 5n5/20 k are needed.

Note from *Figures 10.45* and *10.46* that the VCO operating frequency can be shifted over a narrow range (only a few per cent) via a control voltage applied to pin 2 of the IC. If this voltage is used, pin 2 should be decoupled via C_2, which needs a value roughly double that of C_1.

Parameter	Min.	Typical	Max.
General Operating voltage range Supply current, quiescent Supply current, active (R_L = 20k)	4.75 V	 7 mA 12 mA	9.0 V
Input Input impedance (pin 3) Smallest detectable V_{in} (at I_L = 100 mA)		20k 20 mV r.m.s.	
Output Max. pin 8 voltage Max. pin 8 sink current Fastest on-off cycling rate Off-state output leakage current On-state saturation voltage (at I_L = 100 mA) Output fall time (R_L = 50 Ω) Output rise time (R_L = 50 Ω)	15 V 100 mA	 $f_0/20$ 0.01 μA 0.6 V 30 ns 150 ns	 25 μA 1.0 V
Centre frequency (f_0) Highest f_0 f_0 thermal stability (p.p.m./°C) f_0 stability vs supply voltage	100 kHz	500 kHz 35 ± 60 0.7%/V	 2%/V
Detection bandwidth (BW) Largest BW (% of f_0) at f_0 = 100 kHz Largest BW skew (% of f_0) BW variation with temp., at V_{in} = 300 mV BW variation with V_{supply}, at V_{in} = 300 mV	10%	14% 3% ±0.1%/°C ±2%/V	18% 6%

Figure 10.44 *Main parameter values of the NE567*

The basic *Figure 10.45* and *10.46* circuits can be usefully modified in a number of ways, as shown in *Figures 10.48* to *10.51*. In the *Figure 10.48* circuit the duty cycle or M/S ratio of the generated waveform is fully variable over the 27:1 to 1:27 range via RV_1, as C_1 alternately charges via R_1–D_1 and the left-hand half of RV_1 and discharges via R_1–D_2 and the right-hand half of RV_1 in each operating cycle. The frequency varies only slightly as the M/S ratio is varied.

Figure 10.49 shows how the oscillator can be used to generate quadrature outputs, in which the square-wave outputs of pins 5 and 8 are

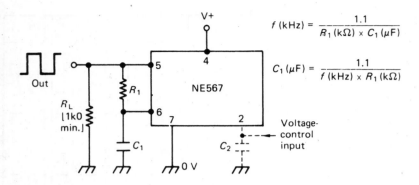

Figure 10.45 *Precision square-wave generator with 20 ns rise and fall times*

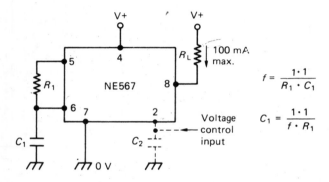

Figure 10.46 *Precision square-wave generator with high-current output*

out of phase by a quarter of a cycle (90°). In this application, input pin 3 is normally grounded; if the pin is biased above 2.8 V, the pin 8 output waveform shifts by 180°.

Finally, *Figures 10.50* and *10.51* shows how the VCO circuits can be modified to enable timing resistor values to be increased to a maximum of about 500 k, thus enabling the C_1 timing value to be proportionately reduced. In both cases, a high-impedance voltage-following buffer stage is wired between the R_1–C_1 junction and pin 6 of the IC. In *Figure 10.50* this buffer takes the form of a simple emitter follower transistor stage, and causes a slight loss of waveform symmetry. The *Figure 10.51* circuit, on the other hand, used an op-amp voltage follower in the buffer position, and this causes no loss of symmetry.

Phase-locked loop circuits 219

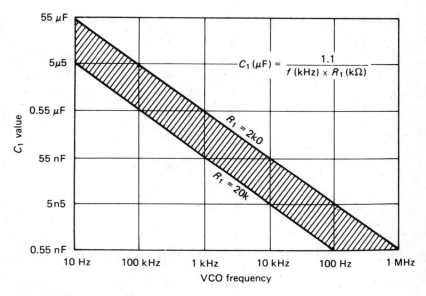

Figure 10.47 *Component-selection guide for the VCO section*

NE567 outputs

The NE567 has a total of five output terminals. Two of these (pins 5 and 6) give access to the VCO output waveforms, and a third (pin 8) functions as the IC's main output terminal, as already described. The

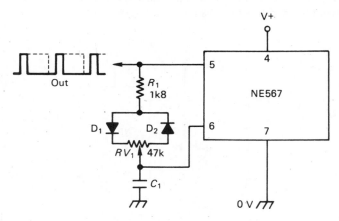

Figure 10.48 *Generator with variable M/S ratio output*

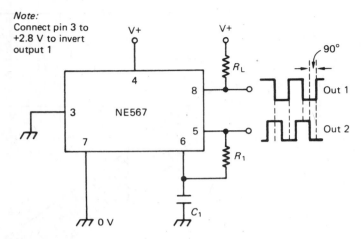

Figure 10.49 *Generator with quadrature outputs*

remaining two outputs are available on pins 2 and 1, and these give the following action:

Pin 2 gives access to the phase detector output terminal of the PLL network, and is internally biased at a quiescent value of 3.8 V. When the IC receives in-band input signals, this voltage varies as a linear function of frequency over the typical range 0.95 to 1.05 f_o (the VCO free-running frequency), with a slope of about 20 mV/% of frequency deviation. *Figure 10.52* illustrates the basic relationship between the pin 2 and pin 8 outputs of the IC when it is used in the tone switch mode, at tone signal bandwidths of 14% and 7%.

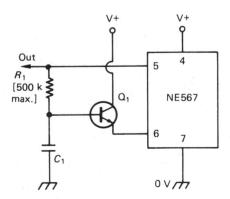

Figure 10.50 *Use of a transistor buffer to increase the permissible value of R_1*

Phase-locked loop circuits 221

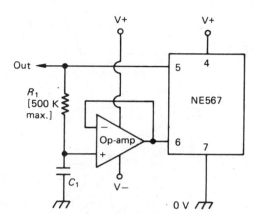

Figure 10.51 *Use of an op-amp buffer to increase the permissible value of R_1*

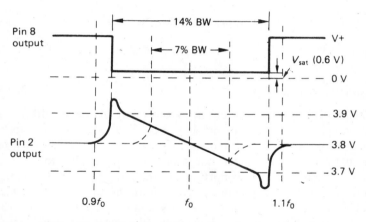

Figure 10.52 *Pin 2 and pin 8 outputs under in-band conditions*

Pin 1 gives access to the output of the ICs quadrature phase detector. During tone lock, the *average* voltage of this pin is a function of the circuit's in-band input signal amplitude as shown in the transfer graph of *Figure 10.53*. Note that the pin 8 output transistor turns on when the pin 1 mean voltage is pulled below the 3.8 V threshold value.

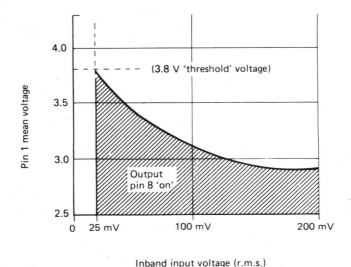

Figure 10.53 *Voltage/signal transfer graph of pin 1*

Bandwidth and skew

When the IC is used in the tone-switch mode, its bandwidth (as a percentage of f_o) has a maximum value of about 14%, but is proportional to the value of in-band signal voltage in the range 25 mV to 200 mV r.m.s. (but is independent of values in the 200 to 300 mV range), and is inversely proportional to the f_o–C_2 product, the actual bandwidth value being equal to 1070 times the square root of $V_{in}/(f_o.C_2)$, where the V_{in} value is in Vr.m.s. and the C_2 value is in microfarads. To select a C_2 value on an empirical (educated trial and error) basis, start by making its value $2 \times C_1$, and then either increase its value to reduce the bandwidth, or reduce it to increase the bandwidth.

Skew is a measure of how well the detection band is centred about the centre frequency (f_o) of the VCO, and is specified as a percentage of f_o by the formula $(f_{max} + f_{min} - 2f_o) 2f_o$, where f_{min} and f_{max} are the frequencies corresponding to the edges of the detection band. Thus, if a tone switch has an f_o of 100 kHz and a bandwidth of 10 kHz, and its edge of band frequencies are symmetrically placed at 95 kHz and 105 kHz, its skew value is 0%. If, on the other hand, its edge of band values are highly non-symmetrical at 100 kHz and 110 kHz, its skew value works out at 5%. In practice, the NE567 gives typical worst-case skew values of 3%. If desired, the skew value can be reduced to zero by feeding an external bias trim voltage to pin 2 of the IC via a pot and a 47k series resistor, as shown in *Figure 10.54*.

Tone-switch design

Practical tone-switch circuits of the *Figure 10.43* type are quite easy to design. Start by selecting the R_1 and C_1 frequency control component values with the aid of the *Figure 10.47* graph, and then select the value of C_2 on the empirical bases described above, starting by making it twice the value of C_1 and then adjusting its value (if necessary) to give the desired signal bandwidth. If band symmetry is critical, skew adjustment (see *Figure 10.54*) can be added at this stage.

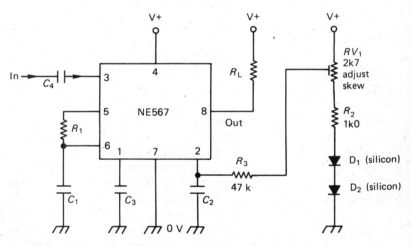

Figure 10.54 *Tone switch with skew adjustment via RV_1*

Finally, to complete the circuit design, give C_3 a value double that of C_2 and check the circuit action. If C_3 is too small the pin 8 output may pulsate during switching transients. The C_3 value can be increased to add switching delays to the pin 8 output terminal.

Multi-switching

Any desired number of NE567 tone switches can be fed from a common input signal, to make a multi-tone switching network of any

224 Timer/Generator Circuits Manual

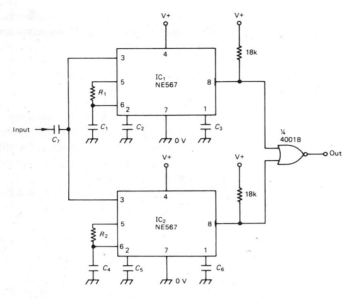

Figure 10.55 *Dual tone decoder with a single output*

desired size. Two particularly useful two-stage switching networks are shown in *Figures 10.55* and *10.56*.

The *Figure 10.55* circuit works as a dual tone decoder, and has a single output that activates in the presence of either of two input tones. The two-tone switches are simply fed from the same signal source and have their outputs NORed via a 4001B CMOS gate.

Finally, *Figure 10.56* shows how two-tone switches can be wired in parallel to act like a single tone switch that has a bandwidth of 24%. In this case the operating frequency of the lower (IC_2) tone switch is simply made 1.12 times lower than that of the upper (IC_1) tone switch, so that their switching bandwidths overlap.

Phase-locked loop circuits

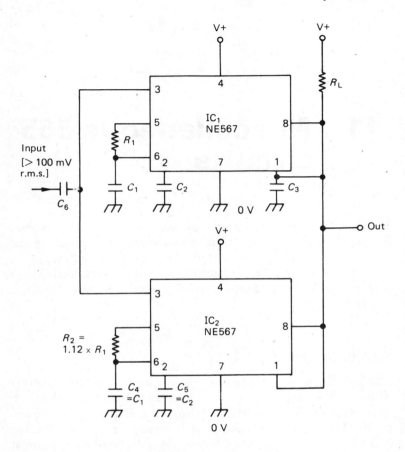

Figure 10.56 *Tone switch with 24% bandwidth*

11 Miscellaneous 555 circuits

In Chapter 5 we explained the basic features of the popular 555 timer IC and showed how it can be used in a variety of monostable and astable multivibrator circuits, and in Chapter 6 we showed a variety of ways of using it as a triggered sawtooth generator. In this final chapter we return to the 555, and show a selection of practical ways of using it in a variety of useful circuits.

Schmitt triggers

The 555 IC can be made to function as a useful Schmitt trigger by wiring its pin 2 and pin 6 *comparator terminals* together, as shown in *Figure 11.1*, and applying the external input signals directly to these two points. These two comparators are internally biased via a built-in potential divider that sets the inverting pin of the upper comparator at $2/3\ V_{cc}$ and the non-inverting pin of the lower one at $1/3\ V_{cc}$. The comparator outputs drive the R and S terminals respectively of the output-driving R–S flip-flop. Consequently, the action of this circuit is as follows:

When the input terminal voltage rises above $2/3\ V_{cc}$ the IC output switches low, and remains there until the input falls below $1/3\ V_{cc}$ at which point the output switches high and remains there until the input rises above $2/3\ V_{cc}$ again. The difference between these two trigger levels is called the *hysteresis value* of the circuit, and has a value of $1/3\ V_{cc}$ in this case. This large hysteresis value makes the circuit useful in noise/ripple-rejecting signal conditioning applications, as indicated in the diagram.

Figure 11.2 shows the above circuit modified for use as a high-

Miscellaneous 555 circuits 227

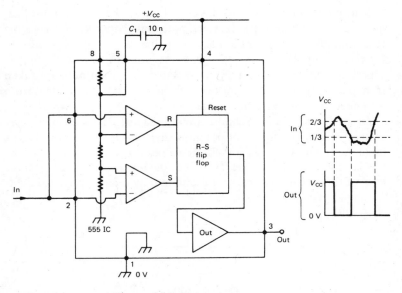

Figure 11.1 *Simple but useful Schmitt trigger*

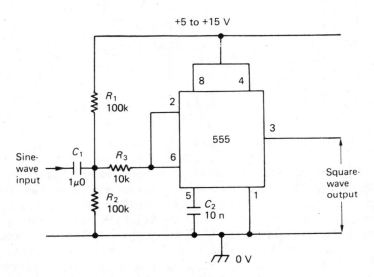

Figure 11.2 *555 Schmitt sine/square converter*

performance sine to square converter that can be used at input frequencies up to about 150 kHz. Here, potential divider R_1–R_2 biases the IC's 2 and 6 input pins at a quiescent value of $1/2\ V_{cc}$ (i.e., mid-way between the upper and lower trigger values), and the sine-wave input is superimposed on this point via C_1. Square-wave outputs are taken from pin 3. R_3 is wired in series with the input signal to isolate it from the effects of the 555's switching actions.

Figure 11.3 shows the Schmitt circuit used as a dark-activated relay switch, with light-dependent potential divider RV_1–LDR wired to its input terminal. The RV_1 and LDR values are roughly equal at the median switching light-level. Note that the inherently large hysteresis level of this circuit may make it useful in only a few specialized light-sensing applications.

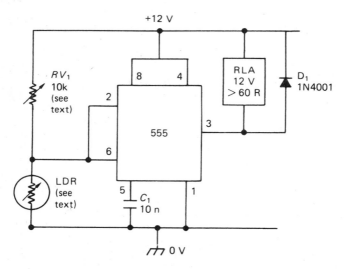

Figure 11.3 *Dark-activated relay switch has substantial hysteresis*

A far more useful relay-driving dark-activated switch is shown in *Figure 11.4*. This circuit acts as a fast comparator rather than as a true Schmitt trigger. It has the input (pin 6) of its upper internal comparator tied high via R_1, and the light-sensing RV_1–LDR potential divider is applied to the input (pin 2) of the lower comparator. The LDR needs a resistance in the range $470R$ to $10\,k$ at the turn-on light level.

Note that the above circuit can be made to act as a light- (rather than dark-) activated switch by simply transposing the RV_1 and LDR pos-

Miscellaneous 555 circuits 229

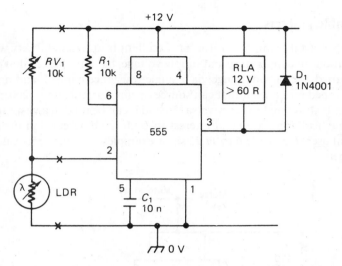

Figure 11.4 *Minimum-backlash dark-activated relay switch*

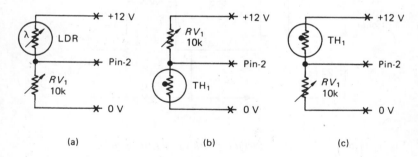

Figure 11.5 *Alternative input circuits for Figure 11.4, to give activation by (a) light, (b) under-temperature, and (c) over-temperature*

itions, as shown in *Figure 11.5(a)*. Alternatively, the circuit can be made to function as a temperature-activated switch by using an NTC thermistor in place of the LDR, as shown in *Figure 11.5(b)* and *(c)*. This thermistor must present a resistance in the range 470R to 10k at the required turn-on temperature.

Astable gadgets

The 555 astable multivibrator has excellent frequency stability with variations in temperature and supply voltage, has a frequency that can be varied over a wide range via a single resistance control, and has a low impedance output that can source or sink currents up to 200 mA. These features make it very versatile, and it can be used in a vast range of practical applications of interest to both the amateur and professional user. *Figures 11.6* to *11.12* show examples of typical 555 astable gadgets.

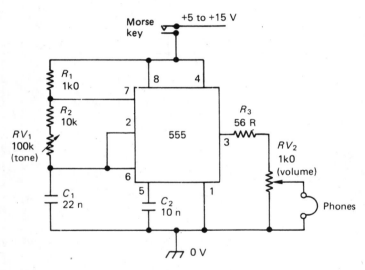

Figure 11.6 *Code-practice oscillator with variable tone and volume*

Figure 11.6 shows the 555 used as a morse-code practice oscillator. The circuit acts as an astable with frequency variable from 300 Hz to 3 kHz via tone control RV_1. The phone volume is variable via RV_2, and the phones can have any impedance from a few ohms upwards. The circuit draws zero quiescent current, since the normally-open morse key is used to connect it to the positive supply rail, which can have any value between 5 and 15 V.

Figure 11.7 shows the 555 used as a simple electronic door-buzzer. The bell-push switch (SW_1) connects the 9 V supply, and the IC's output is fed to a small speaker (25R to 80R) via C_4. C_1 has a low supply-line impedance, ensuring adequate output drive current to the

Miscellaneous 555 circuits   231

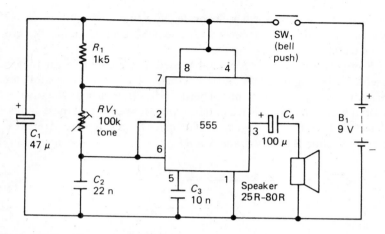

Figure 11.7   *Electronic door-buzzer*

speaker when SW_1 is closed. The circuit generates a monotone buzzer signal.

Figure 11.8 shows the 555 astable used as a continuity tester that generates an audible tone only if the resistance between the test probes is less than a few ohms. Circuit operation relies on the fact that the astable will operate only if pin 4 is positively biased to 600 mV or greater. In the diagram this pin is normally pulled to ground via R_2, so the astable is inoperative.

To enable the *Figure 11.8* astable to operate, the two probes must be closed together, connecting R_2 to the output of the R_3–ZD_1 voltage-reference generator via RV_2. In use, RV_2 is carefully adjusted so that astable operation is barely obtained under this condition. Thus, if the inter-probe resistance exceeds a few ohms when a continuity test is

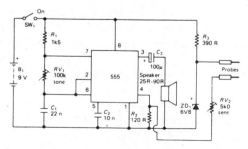

Figure 11.8   *Continuity tester*

being made the astable will not operate. Note that the circuit consumes several milliamperes whenever SW_1 is closed, even if the probes are open circuit.

Figure 11.9 shows the 555 astable used as a signal injector that is useful for testing both AF and RF circuits. The astable operates at a basic frequency of a few hundred hertz when PB_1 is closed. The square-output waveform is very rich in harmonics, however, and these can be detected at frequencies up to tens of megahertz on a radio receiver. The signal injection level is variable via RV_1.

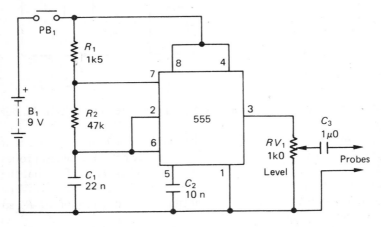

Figure 11.9 *Signal injector*

Figure 11.10 shows the 555 used to make a metronome in which the tick rate is variable from 30 to 120 beats-per-minute via RV_1, and the tick volume is variable via RV_2. This circuit is a modified version of the standard astable, with its main timing network driven from the IC's pin 3 output. When the output switches high C_1 charges rapidly via D_1–R_1 to generate a brief (a few milliseconds) tick pulse. When the output switches low again C_1 discharges via RV_1–R_2, producing an off period of up to 2 s (= 30 beats/minute). The output pulses are fed to a small speaker via volume control RV_2 and buffer Q_1.

LED flashers and alarms

Figures 11.11 to *11.13* show the 555 astable used in LED flasher applications in which the LEDs have equal on and off times. With the

Miscellaneous 555 circuits 233

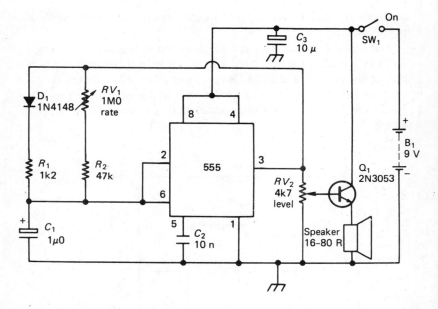

Figure 11.10 *Metronome circuit*

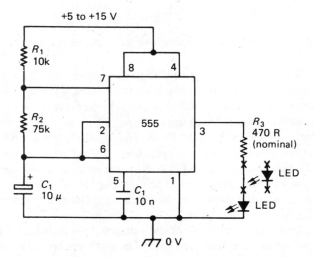

Figure 11.11 *LED flasher with single-ended output*

component values shown the circuits each operate at about one flash per second.

The *Figure 11.11* circuit has a single ended output. Either a single LED or a chain of series-wired LEDs can be put between the IC's output and ground, and all LEDs turn on or off together. R_3 sets the on current of the LEDs. Most LEDs drop about 2 V when on, so several LEDs can be series-wired in a circuit that is powered from a 15 V supply.

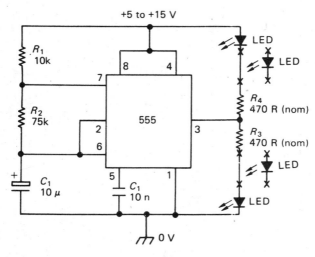

Figure 11.12 *LED flasher with double-ended output connection*

The *Figure 11.12* circuit is similar to the above, but has a double-ended output connection in which all upper LEDs are on when the lower ones are off, and vice versa. R_3 sets the on currents of the lower LEDs and R_4 sets that of the upper ones.

Figure 11.13 shows the basic *Figure 11.11* flasher circuit modified to give automatic dark-activated operation, so that it self-activates at night. In this circuit R_4–R_5–LDR–RV_1 are used as a light-sensitive Wheatstone bridge that is used to activate the 555 astable via balance-detector Q_1 and the pin 4 reset pin of the IC. The circuit operates as follows:

The *Figure 11.13* astable is normally disabled via R_6, which pulls the pin 4 reset terminal to near-zero volts. The astable becomes active only when pin 4 is biased to a positive value of 600 mV or more, and this can

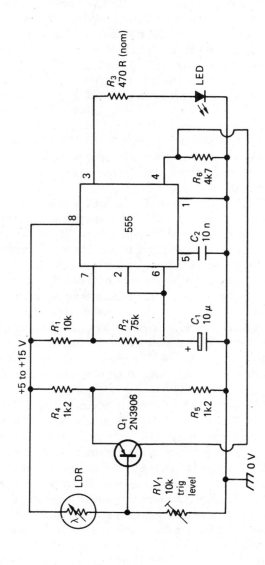

Figure 11.13 *Automatic (dark-activated) LED flasher*

be done by turning on bridge-balance detector Q_1. R_4–R_5 form one arm of the bridge and apply a fixed half-supply voltage to Q_1 emitter, and LDR–RV_1 form the other arm and apply a light-dependent voltage to Q_1 base. Under bright conditions the LDR has a low resistance, so the Q_1 base-emitter junction is reverse biased and the astable is off, but under dark conditions the LDR resistance is high and Q_1 and the astable are biased on. In practice, the LDR must give a resistance in the range 470R to 10k at the dark turn-on level, and RV_1 is adjusted so that the astable just activates under this condition.

The above technique gives precision gating of the 555 astable circuit, and can be used to auto-activate a variety of other 555 astable circuits, to make various audible alarms and relay pulsers, etc. By transposing the LDR and RV_1 positions or replacing the LDR with an NTC thermistor these circuits can be made to auto-active when light or temperature levels go beyond pre-set limits. *Figures 11.14* to *11.16* show practical examples of such circuits.

The *Figure 11.14* circuit gives automatic heat or light activation of a relay pulser, which switches on and off at a once-per-second rate when activated. The relay can be any 12 V type with a coil resistance greater than 60 Ω, and its contacts can be used to activate external electrically-powered devices such as light, sirens, or alarm horns, etc.

The *Figure 11.15* circuit gives automatic heat or light activation of a monotone alarm-call generator, which generates an 800 Hz alarm tone at a power level of several watts in an 8 Ω speaker when activated. Note that the high output current of the circuit may cause modulation of the supply line, so D_1 and C_3 are used to protect the circuitry from the effects of this ripple, and D_2 and D_3 are used to clamp the speaker's inductive switching spikes and thus protect output transistor Q_2 from damage.

Figure 11.16 shows the alternative sensor circuitry that can be used to auto-activate the *Figures 11.14* or *11.15* circuits. For light-sensitive operation the sensor must be an LDR; if activation is wanted when the light level falls to a pre-set dark value the circuit of *Figure 11.16(a)* must be used; if it is wanted when the level rises to a pre-set light value the circuit of *Figure 11.16(b)* must be used. If temperature-sensitive activation is needed, an NTC thermistor must be used as the sensor element. For under-temperature operation, use the circuit of *Figure 11.16(c)*; for over-temperature activation, use *Figure 11.16(d)*. Whatever type of operation is wanted (optical or thermal), the sensor element must have a resistance in the 470R to 10k range at the desired trigger level.

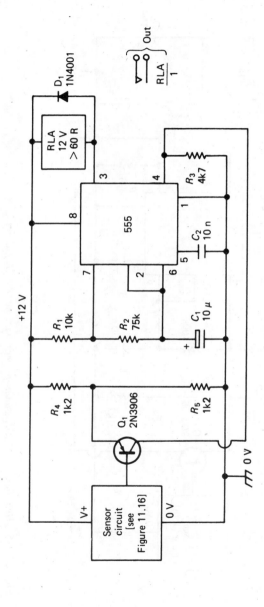

Figure 11.14 *Heat/light-activated relay pulser circuit*

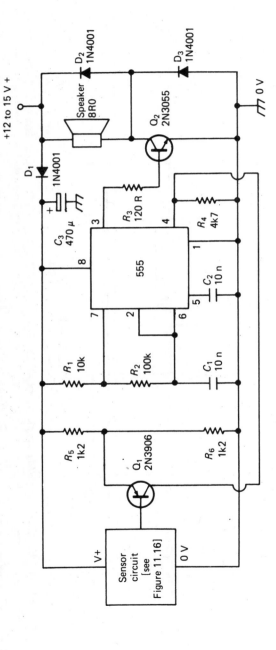

Figure 11.15 *Heat/light-activated medium-power monotone (800 Hz) alarm*

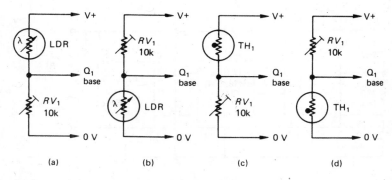

Figure 11.16 *Alternative sensor circuits for use with Figures 11.14 or 11.15, to give activation via (a) dark, (b) light, (c) under-temperature, or (d) over-temperature*

Long-period timers

In Chapter 5 it was shown that a single 555 IC can be used to make an excellent manually-triggered relay-driving timer when connected in the monostable or pulse-generator mode, but that such circuits cannot be used to generate accurate timing periods in excess of a few minutes, since they would require the use of large value electrolytic timing capacitors, which have very wide tolerance values (typically -50% to $+100\%$) and have large and unpredictable leakage currents, and thus cannot give accurate timing.

An easy solution to the problem of obtaining very long but accurate timing periods is shown (in block diagram form) in *Figure 11.17*, which outlines the design of a 60-minute relay-driving timer. Here, the 555 is wired as a 2.28 Hz astable that uses a stable polyester timing capacitor, and its output is fed to the relay driver via a fourteen-stage binary divider that gives an overall division ratio of 16,384. The action of this divider is such that, if its output register is set to zero at the start of the input count, its output will switch high on the arrival of the 8192nd input pulse and will switch low again on the arrival of the 16,382nd pulse, thus completing the count cycle. Thus, the *Figure 11.17* circuit operates as follows:

The timing sequence is initiated by pressing push-button switch PB_1, thus simultaneously connecting the circuit's supply, activating the astable, and (via C_2–R_3) setting the counter to zero count and driving its output low and turning the relay on. As the relay turns on its RLA/1 contacts close and by-pass PB_1, thus maintaining the supply connec-

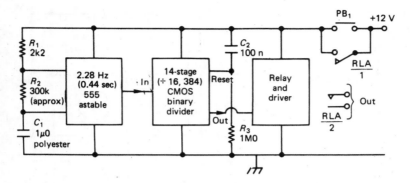

Figure 11.17 *Method of obtaining 60-minute timing period from the 555 IC*

tion once PB_1 is released. This state is maintained until the 8192nd astable pulse arrives, at which point the counter's output switches high and turns the relay off, thus opening contacts RLA/1 and breaking the circuit's supply. The operating cycle is then complete.

Note in this circuit that the astable operates with a period that is only 1/8192nd of the final timing period, i.e., 0.44 s in this case, and that this period can easily be obtained without using an electrolytic timing capacitor.

Figure 11.18 shows how the above technique can be used to make a practical relay-output timer that spans the range 1 to 100 minutes in two overlapping decade ranges. The circuit is powered via a 12 V

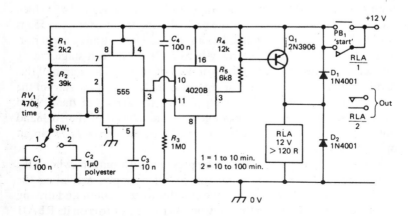

Figure 11.18 *Two-range (1–10 minute and 10–100 minute) relay-output timer*

supply, and the relay can be any 12 V type with two or more sets of change-over contacts and a coil resistance of 120 Ω or greater.

Figure 11.19 shows how the available time delay of the above circuit can be further increased by simply connecting an additional divider stage between the 555 astable output and the input of the relay driver. In this case a divide-by-ten 4017B CMOS IC is wired between the output of the 555 and the input of the 4020B, to give an overall division ratio of 81,920, thus making delays in the range 100 minutes to 20 hours available from this single-range timer. Note that both divider ICs are automatically reset (via C_3–R_3) at the moment of switch-on (PB_1 closure).

Finally, *Figure 11.20* shows how the above circuit can be modified to make a wide-range general-purpose timer that spans 1 minute to 20 hours in three decade-related ranges. Note that the divide-by-ten stage is used on the very-long-period 3 range only.

Analogue frequency meters

A particularly useful application of the 555 is as an analogue meter driver that gives a direct reading of frequency or engine r.p.m. *Figure 11.21* shows a 555 IC used to make a linear-scale analogue frequency meter with a full-scale sensitivity of 1 kHz. The circuit is powered from a stable 6 V supply and needs a pulse or square-wave input with a peak-to-peak amplitude of 2 V or more. Q_1 amplifies this input to a level suitable for triggering the 555 monostable stage, and the pin 3 monostable output is fed to 1 mA FSD moving-coil meter M_1 via multiplier resistor R_5 and offset-cancelling diode D_1. The circuit operates as follows:

Each input waveform cycle triggers the 555 monostable and makes it generate a pulse of fixed amplitude (V_P) and duration (P_P). If the monostable is triggered at a fixed frequency (f), with a period of f_P, the mean voltage (V_M) of the pulse waveform output (integrated over several trigger cycles) is given by:

$$V_M = P_P/f_P \times V_P$$
$$= P_P \times f \times V_P.$$

Thus, if $P_P = 1$ ms and $V_P = 6$ V, it can be seen that $V_M = 3$ V at an input frequency of 500 Hz, or 1.5 V at 250 Hz, or 0.3 V at 50 Hz, etc., and that V_M is directly (linearly) proportional to input frequency. In the *Figure 11.21* circuit the monostable's output pulses are in fact fed to 1 mA

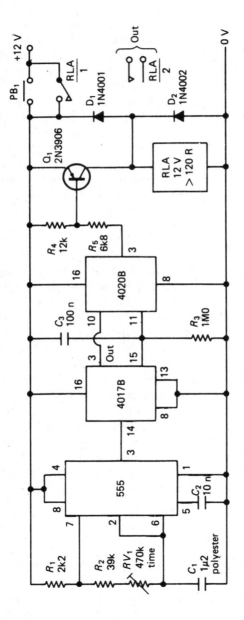

Figure 11.19 *Extra long period (100 minutes to 20 hours) relay-output timer*

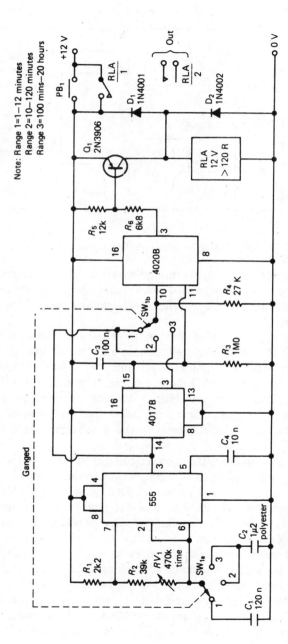

Figure 11.20 *Wide range timer that covers 1 minute to 20 hours in three decade ranges*

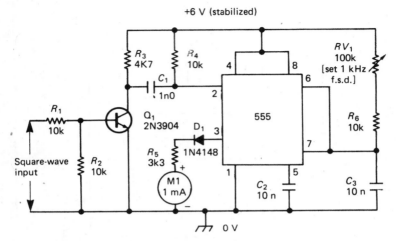

Figure 11.21 *Simple 1 kHz linear-scale analogue frequency meter*

FSD moving coil meter M_1 via multiplier resistor R_5, which gives the meter a full-scale sensitivity of about 3.4 V, and this meter gives a direct reading of the mean pulse-waveform voltage, and thus of the input waveform frequency.

With the component values shown, the *Figure 11.21* circuit is meant to read FSD at 1 kHz. To initially calibrate the circuit, simply feed a 1 kHz square wave into its input and trim RV_1 (which controls the pulse length) to give an FSD meter reading. If desired, the FSD frequency can be varied from 100 Hz to 100 kHz by using alternative C_3 values; the meter can be made to read frequencies up to tens of megahertz by feeding the input signals to the monostable via a digital divider.

Figure 11.22 shows how the above circuit can be modified for use as an analogue tachometer (r.p.m. meter) for use in cars or motor cycles. Here, the circuit is powered by a regulated 8V2 supply derived from the vehicle's battery via R_1–ZD_1–C_1 and the ignition switch, and the monostable is triggered by the vehicle's contact breaker (CB) points via the R_2–C_2–ZD_2 waveform-conditioning network. A 50 µA FSD moving-coil meter is used as the r.p.m. indicator, and is activated from the 555's output via D_1. The circuit action is such that current is applied to the meter via R_5–RV_1 and the IC supply line when the IC output is high, but is reduced to near-zero (via D_1) when the output is low.

The *Figure 11.21* and *11.22* circuits use regulated supply lines, to give a constant pulse amplitude and thus a stable reading accuracy in the

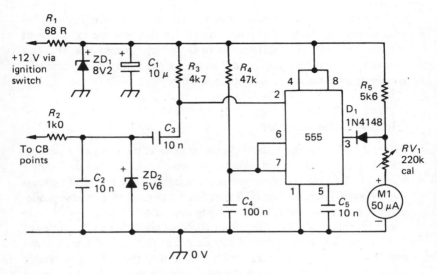

Figure 11.22 *Car/motor-cycle tachometer circuit*

meter. Note, however, that this meter is actually a current-operated device, and in these circuits is used in the voltage indicating mode via the use of suitable multiplier resistors (R_5 in *Figure 11.21*, RV_1–R_5 in *Figure 11.22*. *Figure 11.23* shows (in basic form) an alternative way of

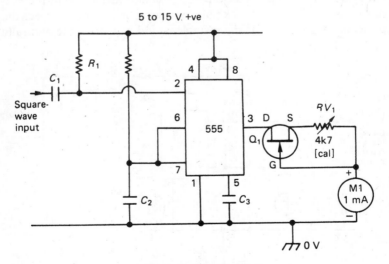

Figure 11.23 *Alternative version of the analogue frequency meter*

making an analogue frequency meter, without the use of either a multiplier resistor or a regulated power supply. In this case the 555's output is fed to the meter via JFET transistor Q_1, which is wired (via RV_1) as a constant-current generator, and thus feeds a fixed-amplitude pulse into the meter irrespective of variations in supply line voltage, etc.

Missing-pulse detector

Figure 11.24 shows how the 555 can be used as the basis of an event-failure alarm or missing-pulse detector, which closes a relay or illuminates a LED if a normally-recurrent event fails to take place. The circuit theory is fairly simple. The IC is wired in the normal monostable mode, except that Q_1 is wired across timing capacitor C_1 and has its base taken to trigger pin 2 of the IC via R_1. This pin is fed with a series of brief pulse- or switch-derived clocking signals from the monitored event, and the R_3–C_1 values are chosen so that the 555's monostable period is slightly longer than the repetition period of the clock input signals.

Thus, each time a clock pulse arrives it rapidly discharges C_1 via Q_1 and simultaneously initiates a monostable timing period that drives pin 3 high, but before that period can terminate naturally a new clock pulse arrives and repeats this process. Consequently, the 555's pin 3 terminal remains high so long as clocking input signals arrive within the prescribed period limit. If a clock pulse is missing or its period exceeds the preset limit, however, the monostable period will terminate naturally

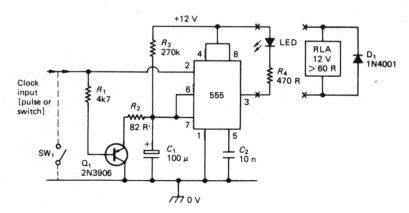

Figure 11.24 *Event-failure alarm or missing-pulse detector with LED or relay output*

and pin 3 will go low and turn on the relay or LED. The circuit thus acts as an event-failure alarm or missing-pulse detector. With the component values shown its natural monostable period is about 30 s, but is variable via R_3–C_1 to suit individual needs.

Voltage converters

The 555 can easily be used to convert an existing d.c. voltage into one of greater value or of reversed polarity, or into an a.c. voltage. *Figures 11.25* to *11.30* show various circuits of such converters.

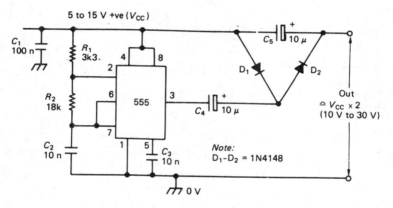

Figure 11.25   *d.c. voltage-doubler circuit*

Figure 11.25 shows the 555 used as a d.c. voltage doubler that gives an output voltage almost twice that of the 555's supply line. Here, the 555 is wired as a free-running 3 kHz astable (with frequency set via R_1–R_2–C_2) that feeds its square-wave output into the C_4–D_1–C_5–D_2 voltage-doubled network, which gives an unloaded output voltage of $2 \times V_{cc}$. Note that C_1 is used to decouple the supply line from 3 kHz ripple, and C_3 is used to further enhance circuit stability.

The *Figure 11.25* circuit can be used with any supply in the range 5 to 15 V and (since it gives a voltage doubler action) can thus be used to give outputs in the range 10 to 30 V. Greater output voltages can be obtained by adding more multiplier stages to the circuit. *Figure 11.26*, for example, shows how to make a d.c. voltage tripler, which can give outputs in the range 15 to 45 V, and *Figure 11.27* shows a d.c. voltage quadrupler, which gives outputs in the range 20 to 60 V.

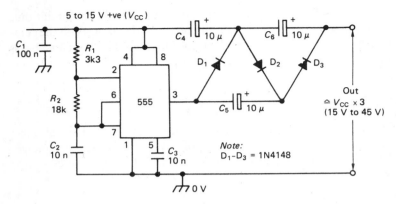

Figure 11.26 d.c. voltage-tripler circuit

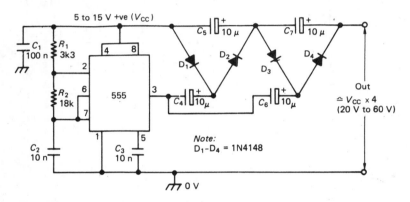

Figure 11.27 d.c. voltage-quadrupler circuit

A particularly useful type of 555 d.c. converter circuit is the negative-voltage generator, which gives an output voltage that is almost equal in value but opposite in polarity to that of the IC supply line. This type of circuit can be used to provide a split supply output for powering op-amps, etc., from a single-ended power source. *Figure 11.28* shows an example of such a circuit, which (like *Figure 11.25*) operates as a 3 kHz astable that drives a voltage-doubler (C_4–D_1–C_5–D_2) output stage.

Another useful type of converter is that which changes a d.c. supply into an isolated (transformer-coupled) a.c. output voltage which can either be used as it stands or can be converted back into a d.c. voltage via a simple rectifier-filter network. *Figures 11.29* and *11.30* show (in basic form) two such circuits.

Miscellaneous 555 circuits 249

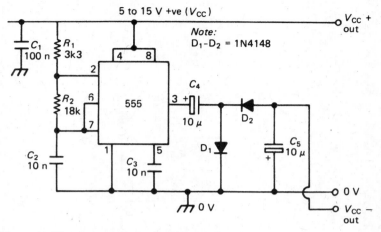

Figure 11.28 *d.c. negative-voltage generator*

The *Figure 11.29* circuit can be used for driving a neon lamp or generating a low-current high-value (up to a few hundred volts) d.c. voltage from a low-value (5 to 15 V) d.c. voltage supply. Here, the square-wave output of the 3 kHz 555 astable is fed into the input of transformer T_1 via R_3. T_1 is a small audio transformer, with a turns ratio sufficient to give the desired voltage, e.g., with a 10 V supply and a 1:20 turns ratio, the unloaded a.c. output is 200 V peak; this a.c. voltage can be converted to d.c. via a half-wave rectifier and filter capacitor, as shown.

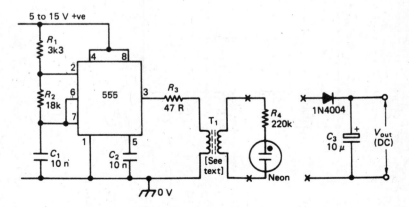

Figure 11.29 *Neon-lamp driver or high-voltage generator*

Miscellaneous 555 circuits / 249

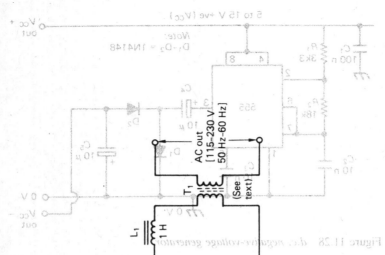

Figure 11.28 d.c. negative-voltage generator

The Figure 11.29 circuit can be used for driving a neon lamp or generating a low-current high-value (up to a few hundred volts) d.c. voltage from a low-voltage (5 to 15 V) d.c. supply. Here, the square-wave output of the 3 kHz 555 astable is fed into the input of transformer T_1 via R_3. T_1 is a small audio transformer, with a turns ratio sufficient to give the desired voltage, e.g., with a 10 V supply and a 1:20 turns ratio, the unloaded a.c. output is 200 V peak; this a.c. voltage can be converted to d.c. via a half-wave rectifier and filter capacitor, as shown.

Figure 11.29 Neon-lamp-driver or high-voltage generator

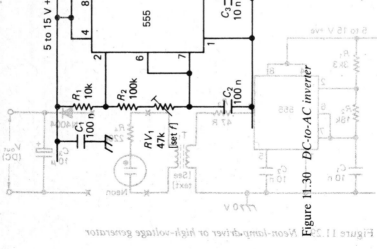

Figure 11.30 DC-to-AC inverter

Finally, the *Figure 11.30* DC-to-AC inverter circuit gives an AC output at mains-line frequency and voltage. The 555 is wired as a 50–60 Hz astable (presettable via RV_1) that feeds its power-boosted (via Q_1–Q_2) output into the input of reverse-connected filament transformer T_1, which has the desired step-up turns ratio. C_4–L_1 acts as a filter that converts the transformer's drive signal into an approximate sine wave.

555 or 7555

To conclude this chapter, here are a few useful notes on the relative merits of the inexpensive bipolar 555 IC and its more expensive CMOS cousin, the 7555.

The standard 555 timer IC is a very popular device, but suffers from a few significant drawbacks. It cannot be used with supplies less than 5 V, and draws a fairly large quiescent supply current (typically 10 mA when operating from 15 V supplies). Worst of all, it draws a brief (a fraction of a microsecond) but massive 400 mA spike of supply current as its output transitions from one state to the other, and this generates an RF noise burst that can play havoc with near-by digital circuits.

The more expensive **7555 CMOS** version of the 555 suffers from none of the above snags. It can use supplies in the range 2 to 18 V, draws only 100 µA of quiescent current from a 15 V supply, and draws a peak spike current of only 10 mA when its output transitions from one state to the other. Also, its *threshold*, *reset* and *trigger* current needs are several orders of magnitude lower than those of the regular 555, enabling timing resistors, etc., to be given values up to hundreds of megohms.

Figure 11.31 shows a rationalized comparative summary of the 7555 and 555 characteristics. Note, on the debit side, that the 7555 performance is inferior in terms of drift-with-voltage accuracy, in pulse-trigger characteristics, and in its output current drive and power dissipation capabilities. Not shown is the fact that the 7555 is also considerably more expensive than the bipolar 555.

Figures 11.32 and *11.33* show, for comparison purposes, the simplified internal circuits of the bipolar (555) and CMOS (7555) versions of the IC. Note particularly the great differences in the resistor values of the two internal voltage divider chains.

The 555 and 7555 ICs are housed in identical 8-pin DIL package styles, and the 7555 can be used as a direct plug-in replacement for the 555 in all circuits shown in this volume. Note, however, that (for

252 Timer/Generator Circuits Manual

Parameter	ICM 7555	Bipolar 555
Power supply range	2V0 to 18 V	4V5 to 16 V
Supply current at V_{CC} = 15 V	0.1 mA	10 mA
Output current, max.	100 mA	200 mA
Power dissipation, max.	200 mW	600 mW
Peak supply current transient	10 mA	400 mA
Timing accuracy, drift with V_{CC}	1%/V	0.1%/V
Input current, trigger	0.01 nA	100 nA
Input current, threshold	0.01 nA	500 nA
Input current, reset	0.02 nA	100 µA
Output rise and fall times	40 ns	100 ns
Minimum trigger-pulse width	90 ns	20 ns

Figure 11.31 *Rationalized comparative-performance summary of the 7555 and 555 ICs*

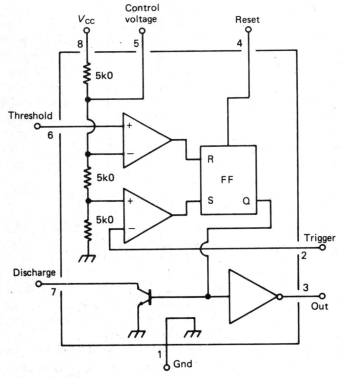

Figure 11.32 *Internal circuit of the bipolar 555 IC*

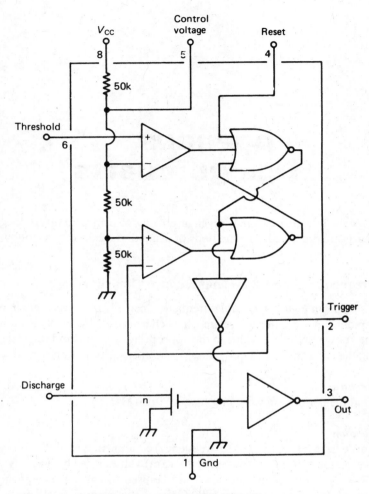

Figure 11.33 *Internal circuit of the CMOS 7555 IC*

reasons of cost) the 7555 should only be used in preference to the 555 in circuits where current economy is important, or where supply voltages are below 5 V, or in complex multi-IC digital circuits where radiation problems might otherwise occur.

Finally, the reader is reminded that dual versions of both the 555 and 7555 are available in 14-pin DIL IC packages. The dual 555 is known as the 556, and the dual 7555 is known as the 7556. *Figure 5.2* shows the IC outline and pin notations that are common to both of these devices.

Appendix
Design charts

To conclude this volume, seven useful waveform generator design charts are presented. The methods of using these charts are as follows.

Frequency-period-wavelength conversion chart

The chart in *Figure A.1* enables frequency, period and wavelength to be rapidly and accurately co-related. The chart is used by simply locating the known parameter value in the appropriate column, and then reading the adjacent equivalent values of the alternative parameters.

Example
Find the period and wavelength equal to 20 MHz.

Solution
Locate 20 MHz in the centre (frequency) column, and read the equivalent period (50 ns) and wavelength (15 m) values directly. The key of *Figure A.1* shows how the chart can be used to cover the frequency range 0.001 Hz (1 mHz) to 1000 GHz, with appropriate changes in the period and wavelength designations.

Symmetrical twin-T or Wien bridge component selection charts

The charts in *Figures A.2* and *A.3* enable the R and C values of a symmetrical twin-T or Wien bridge oscillator or filter network of known frequency (or the frequency for known R and C values) to be rapidly and accurately determined. *Figure A.2* enables the approximate

Appendix Design charts 255

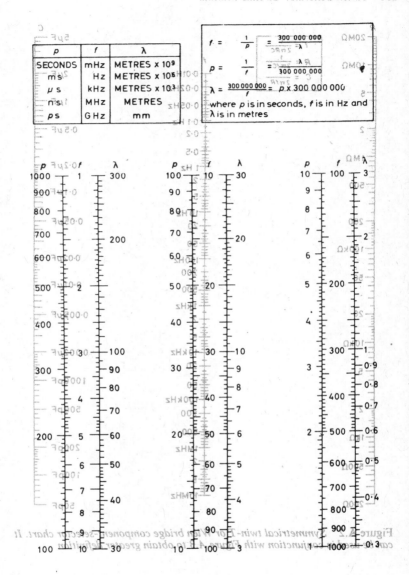

Figure A.1 *Frequency-period-wavelength conversion chart*

256 Timer/Generator Circuits Manual

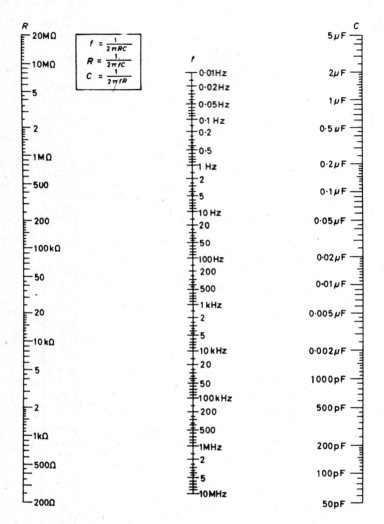

Figure A.2 *Symmetrical twin-T or Wien bridge component-selector chart. It can be used in conjunction with Figure A.3 to obtain greater definition*

parameter values to be determined, and *Figure A.3* enables these approximate values to be translated into precise ones. The charts are used by simply laying a ruler or perspex straight edge so that it cuts two known parameter values, and then reading the value of the third parameter at the point where its scales are cut.

Appendix Design charts 257

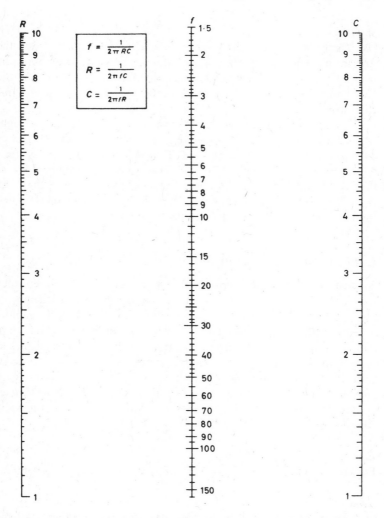

Figure A.3 *Symmetrical twin-T or Wien bridge component-selection chart with expanded scale, to be used in conjunction with Figure A.2. All numerical values can be increased or decreased in decade multiples*

Example

A 1 kHz Wien bridge oscillator is to be made using C values that are decade multiples or submultiples of 1, and with R values that are greater than 2k0 but less than 100 k. Find the required R and C values.

Solution

Use *Figure A.2* to find the approximate R and C values. Pivot the straight edge on the 1 kHz point of the centre scale, and rotate the straight edge so that it sweeps between the designated R limits until a qualifying C value is located. Read off the C (0.01 μF) and R (about 15 k) values.

Use *Figure A.3* to find the precise R values required. Lay the straight edge so that it cuts the 10 value (equal to 1 kHz) of the frequency column and the 10 value (equal to 0.01 μF) of the C column, and read off the R value of 1.59 (equal to 15.9 kΩ) at which the straight edge cuts the R column. Note that all numeric values of the *Figure A.3* chart can be increased or decreased in decade multiples. Hence the values needed to make the 1 kHz Wien bridge oscillator are 0.01 μF and 15.9 kΩ.

L-C tuned circuit component selection chart

Two charts are presented in *Figures A.4* and *A.5*. Between them they enable the C and L values of a tuned circuit of a known frequency (or the frequency resulting from known values of C and L) to be rapidly and accurately determined. The two charts are intended to be used in conjunction. *Figure A.4* enables the approximate parameter values to be determined, and *Figure A.5* enables these approximate values to be translated into precise ones. The charts are used by simply laying a ruler or perspex straight edge so that it cuts two known parameter values, and then reading the value of the third parameter at the point where its scales are cut.

Example

Find the resonant frequency of a 200 μH coil and a 100 pF capacitance.

Solution

Use *Figure A.4* to find the approximate frequency value. Lay the straight edge so that it cuts the 200 μH and 100 pF values, and read the frequency value as approximately 1.1 MHz.

Use *Figure A.5* to find the precise frequency value. Lay the straight edge so that it cuts the 2 value (equal to 200 μH) of the L column, and the 1 or 10 value (equal to 100 pF) of the C column and read off the values of 11.25/3.57 or 35.7/11.25 on the frequency column. Clearly, the 11.25 value (equal to 1.125 MHz) is the correct one that approximates 1.1 MHz. Note that all numeric values of the *Figure A.5* chart

Appendix Design charts 259

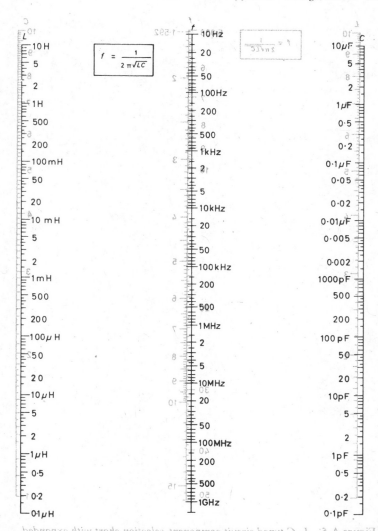

Figure A.4 *L–C tuned circuit component-selection chart. It can be used in conjunction with Figure A.5 to obtain greater definition*

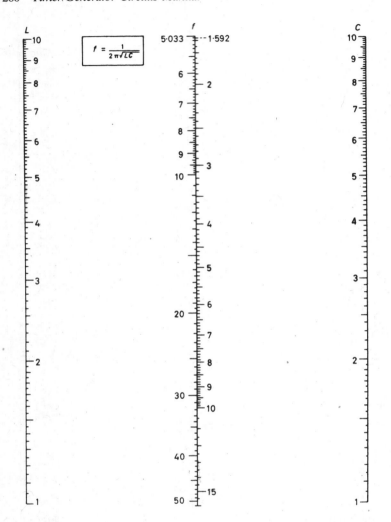

Figure A.5 *L–C tuned circuit component-selection chart with expanded scale, to be used in conjunction with Figure A.4. All numerical values can be increased or decreased in decade multiples*

can be increased or decreased in decade multiples in conjunction with the results of the *Figure A.4* chart.

Timing component selector for transistor or CMOS symmetrical astable multivibrators

Figure A.6 enables the necessary R and C timing component values of a transistor or CMOS symmetrical astable multivibrator (or any other oscillator/astable in which $f = 1/(1.4\ R \cdot C)$ to be rapidly determined. The chart is used by laying a straight edge so that it cuts two known parameter values, and then reading the value of the third parameter at the point where the scales are cut.

Example

A 2 kHz transistor symmetrical astable multivibrator is required. Its R value must be within the range 10 k to 100 k, and its C value must be a decade submultiple of 1. Find suitable R and C values.

Solution

Pivot the straight edge on the 2 kHz point of the centre scale, and rotate the straight edge so that it sweeps between the designated R limits until a qualifying C value is located. Read off the C (0.01 μF) and R (approximately 36 k) values.

Timing component selector for transistor or CMOS monostable multivibrators

Figure A.7 enables the necessary R and C timing component values of a transistor or CMOS monostable multivibrator (or any other monostable in which $P = 0.7\ R.C$) to be rapidly determined. The chart is used by laying a straight edge so that it cuts two known parameter values, and then reading the value of the third parameter at the point where the scales are cut.

Example

A transistor monostable multivibrator is required to produce output pulse widths of 100 μs. Its R value must be within the range 10 k to 100 k, and its C value must be a decade submultiple of 1. Find suitable R and C values.

262 Timer/Generator Circuits Manual

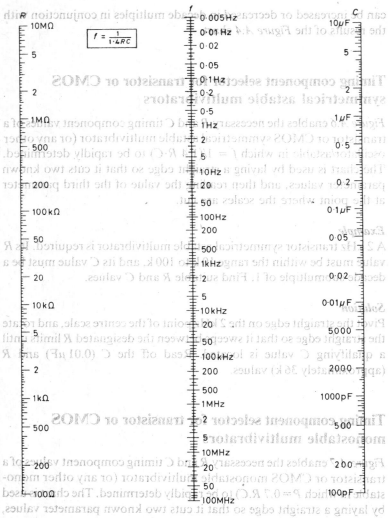

Figure A.6 *Timing component selector for transistor or CMOS symmetrical astable multivibrators*

Solution
Pivot the straight edge on the 100 μs point of the centre scale, and rotate the straight edge so that it sweeps between the designated R limits until a qualifying C value is located. Read off the C (0.01 μF) and R (approximately 14 k) values.

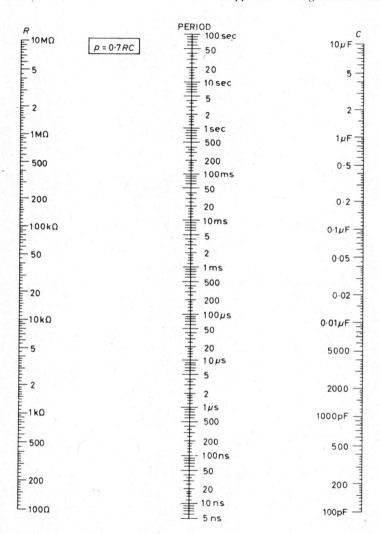

Figure A.7 *Timing component selector for transistor or CMOS monostable multivibrators*

Index

Alarms, 113, 232–8, 246
Amplitude modulation (AM), 152–3
Analogue frequency meters, 241–6
Astable:
 circuit modifications, 34–5
 CMOS basics, 44–7
 CMOS Schmitt, 54–6
 gadgets, 230–2
 gated CMOS, 50–2
 gating, 108
 operation (555), 103–5
 precision, 108
 ring-of-three, 52
 sirens/alarms, 113–16
 transistor circuits, 30–3
 TTL Schmitt, 59–60
 variations, 47–9
Audio oscillator, 163

Car (timer circuits), 94–7
Circuit:
 modifications, 34–5
 resettable, 75–6
 variations, 24–6, 42–4
Clapp oscillator, 26
CMOS:
 astable basics, 44–7
 astable circuits (gated), 50–2
 astable Schmitt, 54–6

bistable circuits, 60–2
monostable circuits:
 4001B, 4011B, 73–5
 4047B, 4098B, 79–83
 flip-flop, 76–9
Colpitts oscillator, 26
Component selection, 256–63
Continuity tester, 231
Conversion chart, 255
C–R:
 oscillator circuits, 14–16
 time constants, 164–70
Crystal oscillators, 174–7

Design charts, 254
Diode-stabilization, 19–21
Door buzzer, 231

Edge detector circuits, 65–8

Feedback oscillator, tuned collector, 23
Flashers, 232
Frequency:
 demodulator, 206–7
 multiplication, 9
 shift keyed, 151–2
 synthesis, 10

Gouriet oscillator, 26

Index

Half mono:
   pulse generator, 65
   stable generator, 65

Hartley oscillator, 24–5
High frequency synthesis, 11

L–C oscillator circuits, 22–5
LED flashers/alarms, 232–8
Linear staircase, 177–9
Lock detector/indicator, 196–8

Metronome circuit, 233
Missing-pulse detector, 246–7
Monostables:
   flip-flop, 76
   half, 65
   resettable, 75
   retriggerable, 78
Morse-code oscillator, 230–1

Op-amp:
   circuit variations, 42–4
   circuits, 126–9, 134–7
   square wave generators, 35–8
Oscilloscope timebase, 132–3

Phase-locked loop (PLL), 8–10, 182–225
Phase-shift oscillator, 15
Pink noise generation, 172–3
Porchlight, 97–9
Pulse generator:
   basics, 63–5
   circuits, 63–86, 99–104, 150–1
Pulse width modulation (PWM), 7

Reinartz oscillator, 26
Relay pulse circuit, 236
Reset-pulse, 68
Resettable circuits, 75–6
Resistance activation, 40–2
Ring-of-three astable, 52–4

Sawtooth generators, 124–33
Schmitt triggers, 226–9
Signal injector, 232
Signetics Corporation, 201
Sine-to-square converters, 30
Sinewave:
   distortion, 157–8
   generators, 14–29, 145–7
   synthesizer (digital), 179–81
Sirens, 113–15, 190–1
Sound effects, 188–94
Square wave generators, 30–61, 147–50
Switch, noiseless press button, 68
Switching, multi, 223–5
Symmetry, variable, 38–40
Synthesis, 10–11
Synthesizer, digital sine wave, 179–81

Tachometer circuit, 245–6
Thermistor stabilization, 17–18
Time constants, 164–70
Timers:
   long period, 239–43
   precision (2N1034), 117–21
   programmable (2240IC), 121–3
Tone switch, 223
Transistor:
   astables, 30–3
   monostables, 69–72
Triangle generators, 124–33, 147–50
Triggered pulse/sawtooth generator, 140–1
Triggering, electronic, 72–3
TTL Schmitt astable circuits, 59–60
Tuned collector feedback, 23–4
Twin-T oscillators, 21–2

Unijunction:
   circuits, 138–40
   transistor (UJT), 124

Voltage controller oscillator (VCO), 7–13
  circuit (4046B), 56–8, 182–8, 194–6
  design, 216–19
Voltage converters, 247–51

Waveforms:
  crystal generated, 6
  free running, 1–3
  generating ICs, 7
  modifying, 164
  modulation, 6, 26–9
  multi, 134–41
  pink noise, 6, 172–3
  ramp, 5
  sawtooth, 3, 124
  sinewave, 14
  square wave, 3–5
  synthesizer ICs, 142–63
  staircase, 5
  triangle, 5, 124
  triggered, 3
  white noise, 6, 170–2
White noise generation, 170–2
Wienbridge oscillators, 16–21

Circuits by type number

555, 7, 87, 117, 129–33, 226–53
556, 87
2240, 87, 121–3
4001B, 50, 60, 73–5
4011B, 50, 60, 73–5
4017B, 200
4027B, 78
4046B, 56–9, 182–8, 194–6
4047B, 79–83
4093B, 54–5, 60
4098B, 79–83
4518B, 200
4528B, 79
7555, 87–9, 251–3
7556, 87–8
40106B, 54
74121N, 83–6
ICL 8038, 158–63
NE565, 202–9
NE566, 208–13
NE567, 214–18
XR2206, 142–55
ZN1034, 87, 117–21